TEXAS

WITH

PARTICULAR REFERENCE TO GERMAN IMMIGRATION
&
THE FLORA, FAUNA, LAND AND INHABITANTS

BY DR. FERDINAND ROEMER

TRANSLATED FROM THE GERMAN
BY OSWALD MUELLER

Copano Bay Press
2011

Originally published in 1849 at Bonn under the title *Mit besonderer Rücksicht auf deutsche Auswanderung und die physischen Verhältnisse des Landes nach eigener Beobachtung*

CONTENTS

Publisher's Note

I don't think that I can express emphatically enough the joy associated with producing an edition of Dr. Roemer's fine work. His astute observations, fearless attitude, respect for the land in which he was a guest and diligent note-taking gave us an uncharacteristically readable chronicle of our land at a pivotal time in our history. When he arrived in Texas, it was a Republic. When he departed, it was one of the United States. The Civil War had not yet scarred the nation. The nature of Texans as observed in this text seem to not have changed much. Dr. Roemer was, indeed, a remarkable man who gave us a timeless gift.

With regard to the text, I am pleased to report that no substantial changes have been made from Oswald Mueller's original excellent translation from the German. A word or two, here and there, were changed when it was obvious that a long-ago editor had fallen asleep at the wheel. Beyond that, the text is true to Mueller's translation, including all scientific names.

An index of scientific names, as well as an index of places and watercourses have been added to this edition. Because we have chosen to retain and expand a bit on Roemer's original descriptive chapter headings to help guide the reader, and because the author rarely used full names of the individuals he referenced in the text, a complete index was not in order.

The modern reader should be aware that some confusion may arise in the casual reading of this text, owing to Dr. Roemer's occasional use of the label "native American." He is referring, of course, to Anglo-Americans born in the United States, rather than to the people who inhabited the land prior to the arrival of Europeans. The definition we have since assigned to the same words in modern ethnic parlance do not apply here, nor have corrections to the text been made to modernize this language.

To Mr. Jonathan R. Peters, whose curiosity, strength of character and persistence rival that of the esteemed Dr. Roemer and men of his ilk, this book is dedicated with unceasing love and the greatest affection.

-Michelle M. Haas, Managing Editor
Windy Hill

A Preface on the Translation of *Texas*

Prompted by a critique of Dr. Ferdinand von Roemer's incomparable book, *Texas*, as well as an excellent biographical sketch of Dr. Roemer written by Prof. W. S. Geiser in the *Southwest Review*, I undertook the translation of the book.

I wish to extend my sincere thanks to the Rice Institute of Houston, Texas, for so generously lending me a copy of the book. I desire to acknowledge gratefully the many helpful suggestions received from Miss Olive Rachel Hester and Miss Jimmie May Hicks, as well as to express my sincere thanks to my brother, A. B. Mueller, who helped me so materially. Furthermore, I wish to make mention of the valuable help received from Dr. Donald C. Barton and Dr. Robert Blau in the translation of the geological terminology, and to Carl W. Sauer, to whom I am indebted for the cover design.

To the reader unacquainted with the German language, the length of certain sentences in this volume may appear extraordinary. This is due to the desire of the translator to preserve, insofar as clarity permits, the flavor of von Roemer's style. The translator also desires to call attention to the antiquated spelling of various proper names. These do not follow the modern spelling, but that of von Roemer's day.

While the question of the adequacy of the translation must be left to the decision of the reader, the translator has, nevertheless, derived keen pleasure from his task of transferring the treasures of the work from the original German to the present day English.

May this book, as a contribution to the Centennial, stimulate in us a greater appreciation of the trials and hardships endured by our forefathers in the founding of this great State — Texas.

— Oswald Mueller

GEOLOGICAL PREFACE

Ferdinand Roemer commonly is spoken of as "the father of Texas geology." William Kennedy in 1841 published an extended treatise on Texas and gave in it a careful account of the scanty and fragmentary knowledge of the geology of Texas as it was known at that time. Scherpf in 1841 and Solms-Braunfels in 1846 published maps with observations in regard to the then known geology of Texas. But Roemer was the first well-trained geologist to make an extensive and detailed study of Texas geology. His work, *Cretaceous Formations of Texas and Their Organic Inclusions (Die Kreidebildungen von Texas und ihre Organischen Einschluesse*, Bonn, Adolphus Marcus, 1852) was the first monograph on Texas geology. The present volume, *Texas*, and a short paper, give information in regard to other features of the geology, mainly the stratigraphy of Texas. In these papers he gives a general description of the "Azoic," Paleozoic, Cretaceous, Tertiary and Quaternary rocks, and on his map gives the general boundaries of the Tertiary, Cretaceous and Paleozoic beds in central Texas.

Much of his geological observation and reasoning is just as good now and always will be as good as it was when he made them. But his work was done in a very early day geology, for the modern era of geology began only with the beginning of the nineteenth century, and since his work was in the 1840s, many geologists have studied the same formation.

As such, much of his geological observation therefore has to be reinterpreted in the light both of modern geologic knowledge and of our much more detailed and extensive knowledge of the geology and stratigraphy of the area which he studied. The lay reader therefore should be cautious in accepting Roemer's interpretation of the geology.

Roemer's explanation of the springs at New Braunfels and elsewhere along the eastern foot of the Edwards Plateau for example now is known not to be correct. Observation since Roemer's day has shown that the Upper Cretaceous clays of the "Black Prairie" in front of the Balcones Escarpment are younger than the Edwards limestone and other limestones of the plateau and not older. Those clay formations do not extend under the limestones as Roemer thought. The Balcones Escarpment represents what is called a "fault," a break in the earth's crust along which the clay formations to the east were dropped down in reference to the limestones of the plateau. Rain water falling on the plateau seeps down the dip of the beds of the plateau to the fault and then comes up the fault plane to the surface to give the springs at New Braunfels, San Marcos and Austin.

In explaining the setting of the building stone at San Antonio, the modern geologist would say that the water in the pore spaces of the rock contained lime or other minerals in solution and that the evaporation of the water caused the precipitation of those minerals and therefore produced cementation of the constituent grains and the "setting" of the rock which Roemer describes.

The term "Transition" which Roemer uses frequently in connection with rocks adjacent to the Central Mineral Region harkens back to an archaic terminology and theory in historical geology. Curiously the modern geologist still uses part of that old terminology in his "Tertiary" and "Quaternary," but they have wholly lost their ancient meanings.

— DONALD C. BARTON
HOUSTON, TEXAS
JULY 12, 1935

FOREWORD

When in the late fall of 1847 I returned to Germany from a trip to North America, lasting several years and which was primarily made for the purpose of geological studies, it was at first my intention to give a general report embracing at the same time my entire trip, reserving the purely scientific data for a special work. In the meantime I soon became convinced that the portion dealing with my sojourn in Texas, was for various reasons, better suited for a separate report. Europe had but meager information concerning this part of North America and since my relatively longer stay in Texas afforded opportunity for somewhat closer and more extensive observations, it seemed advisable to give a more detailed report on this section. The author also felt somewhat duty-bound, considering the German emigrants, to publish his humble contribution toward a better understanding of this country more quickly than would have been possible had it been included as a part of a more elaborate work, as at first planned. Thus the present book on Texas, based on observations made during my stay, December 1845 to April 1847, came into being.

The principal part of the book contains a continuous travelog of my sojourn in Texas. In the introduction, reference is made to various general topics which could not well be included in the travelog. In particular were the circumstances regarding German emigration more frequently discussed.

The geognostical map accompanying this book is intended to serve at the same time as a geographical map for the reader not interested in geology; nor should the simple light color, with which the various rock formations are designated, offer any real obstacle.

The author takes this opportunity to discharge the pleasant duty of thanking those who assisted him in carrying out the purpose of his trip. Most readily and cheerfully does he acknowledge that only through the effective interest in his undertaking, manifested in various ways, was its execution to the desired extent possible. Also, the Royal Academy of Berlin deemed the author's scientific endeavors worthy of support. I shall always feel deeply grateful to Baron O. von Meusebach who, during my entire sojourn in Texas, showed such friendly interest in general and at the same time used his position as Commissioner-General of the Mainzer Verein to further my cause in every way. Also to the former officials of the Verein at New Braunfels and Fredericksburg (especiallly Mr. v. Coll who, during the absence of v. Meusebach, managed the affairs of the Verein at New Braunfels,) I want to express my sincere thanks for their friendly good will and their unselfish obligingness, of which they continually gave me evidence.

— THE AUTHOR
POPPELSDORF AT BONN
AUGUST 1849

SECTION I
LOCATION & BOUNDARY OF TEXAS

Texas, originally a part of the vice-regency of New Spain, later a province of the Mexican Republic, then an independent republic, and since 1846 a State of the Republic of the United States, comprises a vast area on the Gulf of Mexico between the Mississippi River basin in the northeast and the Rio Bravo del Norte or Rio Grande in the southwest. It extends from latitude 26° to 38° N.L.

The boundaries of the country established by a resolution of the Congress of the Republic of Texas in 1836 are formed in the east and north (in addition to several artificial boundaries) by the Sabine River, the Red River and the Arkansas; in the west and southwest by the Rio Bravo del Norte, or the Rio Grande; and in the southeast by the Gulf of Mexico.

With reference to the western boundary it should be noted, however, that the establishment of the upper course of the Rio Grande as such by the Republic of Texas appears altogether arbitrary, nor has it at any time attained practical significance. According to this arrangement, the larger and more populous part of the province of New Mexico, including the capital, Santa Fe, would belong to Texas. This province had no political connection with Texas at any time, nor did military operation of the Anglo-American settlers during the War of Independence with Mexico extend as far as these distant regions. The few abortive expeditions by several hundred adventurers, equipped in Austin in 1841, are not regarded as such. The natural boundaries also decidedly separate Texas proper from the upper valley of the Rio Grande, since a broad highland, several hundred miles long, extends between them. At the present time it is inhabited by roving Indian bands, and no doubt the greater part of it will never be accessible for cultivation by civilized man. The inhabited highland, in which some of the largest rivers of the country have their source, forms the natural western boundary of Texas. Only in its lower course, from Presidio Rio Grande to the point where it empties into the sea, does the Rio Grande form a natural boundary of Texas in the southwest.

Through a recently signed treaty of peace by which New Mexico was acquired by the United States, the establishment of the western boundary has now become an affair of the North American Republic instead of an international one, and its solution should offer no serious difficulty. Through a bill introduced in Congress in January 1849, whereby the springs of the Red River and the Arkansas are set as the eastern boundary, the settlement of the matter has already begun.

SECTION II
PHYSICAL APPEARANCE OF THE COUNTRY

Texas is divided into three zones according to its surface, which are readily distinguishable from each other and in most places form a clear line of demarcation.

(1) The first zone is the low, level region which stretches along the entire coast from the Sabine to the Rio Grande and from these points extends more or less into the interior. The width of this level coast country varies. While it is only about thirty miles wide at the Sabine, it reaches a width of a hundred miles at the Colorado and then decreases gradually as it nears the Nueces. The elevation of this low coastal plain also varies. At the lower end of Galveston Bay it is scarcely a few feet above sea level, while the prairie at Houston rises to an elevation of sixty to seventy feet above sea level. The greatest portion of this level coast country is composed of open prairie broken only by narrow fringes of forest along the rivers and creeks. In part these prairies are very fertile but, on account of being so low, they are usually very wet in the spring and can be traversed only with difficulty, even on horseback. The rich bottom land of Caney Creek and the Bra-

zos River, which is especially adapted for raising sugarcane but which is for the greater part covered with impenetrable canebrakes, also belongs in this zone of the lower coast land.

And finally, also the islands which stretch along the entire coast of Texas as long, narrow strips of land (due to their levelness and low altitude) partake of the nature of the low coast land, although in fertility they cannot compare with the main land. All are barren and sandy, and, with the exception of a small part of Galveston Island, uncultivated. Adjoining the low coastal country, often connected with it by a gradual transition is the second zone:

(II) The gently rolling or undulating region. This comprises the largest and most beautiful region of the cultivated parts of Texas. It extends from the low coastal country to the highlands in the northwest. Its northwest boundary is clearly marked by the farthest settlements and only during the last few years have several German settlements penetrated beyond the rocky highland. This strip of land between the Brazos, Colorado and Guadalupe in a northwest by northeast direction, reaches a width of from one hundred fifty to two hundred miles.

The greater portion of this region is fertile and suited for farming. In the eastern part of Texas from the Red River to the Trinity, this undulating hill country is generally covered with dense forests and its general characteristics resemble the Southwestern States, particularly the adjoining states of Louisiana and Arkansas. In the central section, however, from the Trinity to the Guadalupe, prairies and forests, particularly oak forests, alternate in a fashion which makes this region of Texas so charming. Farther west, beyond the Guadalupe and the San Antonio River, the open prairies predominate and the forests, which are in the main confined to small strips along the water courses, become so scarce, that in many places the establishment of settlements is made difficult, if not impossible altogether.

The land between the Nueces and the Rio Grande has still another aspect. The whole wide area between the lower course of both rivers is an arid, unfruitful wilderness, over-grown with a peculiar vegetation of barren, thorny shrubs, and is inhabited only by herds of mustangs or wild horses and small bands of roving Indians. In its entire course of several hundred miles, extending from Presidio Rio Grande to its mouth, the Rio Grande has no tributaries on its left or northern bank, a circumstance which plainly indicates the lack of rainfall in the adjoining region. This wilderness, where no settlements will ever be made, furnishes almost a more natural boundary between Texas and Mexico than the Rio Grande itself. It was disputed territory up to the time when the United States and Mexico signed a treaty of peace, since both laid claim to it. When the United States troops crossed the Nueces, this was regarded by the Mexicans as the beginning of hostilities.

The last natural division in the geography of Texas is a mountainous region, best described thusly:

(III) The mountainous region, in part rocky, arises still farther from the coast of the Gulf of Mexico beyond the rolling and undulating highland.

The boundary along the highland runs in a general north-easterly direction. Beginning at Presidio on the Rio Grande, it first follows the old Presidio road, then crossing the San Antonio River above San Antonio de Bexar, the Guadalupe at the German settlement at New Braunfels, the Colorado at Austin, the Brazos at is falls, and the Trinity near the con-fluence of its main forks, it continues in the same direction until it reaches the Red River. In some places the boundary is very distinct in contrast to the undulating land. This is the case at New Braunfels, for example, where the physical character of the land changes suddenly upon descending the precipitous slopes of the range of hills north of the city.

The mountain region where all important rivers—the Nueces, the Guadalupe, the Colorado, the Brazos and the Trinity—have their source is little known, since wild ma-rauding Indian tribes and natural obstacles of the terrain make a visit to this region difficult. Higher mountain ranges are not found in this region; these appear rather near the upper course of the Rio Grande. The so-called San Saba Mountains which are supposed to be the highest range in Texas, have been proven non-existent by an expedition of Germans sent out in the spring of 1847. More will be said about this later in the pages of this book. The region in the vicinity of the San Saba River is nothing but a plateau, scarcely two thousand feet above the level of the Gulf of Mexico. It is rocky, barren and sterile and contains only narrow valleys with steep, rocky slopes. Here and there these valleys enlarge themselves into fertile bottoms. While the hills are all about on the same level, the unevenness of the surface is caused by the deep valleys and ravines. A real mountain range is not present, since not even slight elevations above the general level are noticeable. The land on the western bank of the Colorado between its tributaries—San Saba, Llano and Pedernales—as well as south of the Pedernales to the descent of the tableland toward the undulating hilly country between San Antonio, New Braunfels and Austin is of the same character, with the exception that here and there, particularly at the Pedernales, the valleys enlarge themselves and their approaches are less steep and rocky. The highland north of the boundaries, mentioned previously, appears to be similar in character, as far as I know.

The extreme northwestern portion of Texas lying toward the upper course of the Red River and in the direction of the Arkansas, however, shares the natural characteristics of the immense prairies which extend north between the Mississippi and the Rocky Mountains. Wide, immeasurable, almost level prairies, rich in grass, but lacking in wood and water predominate here. They are in part sandy.

In this section are also found the Cross Timbers, those peculiar, continuous, forested hilly zones, several miles wide running north and south, which contrast so sharply with the surrounding prairies. The species of trees of which they are composed are primarily the various kinds of oaks, according to Gregg's *Commerce of the Prairies*, many of which are gnarled and stunted in growth, due in part to the many prairie fires to which they are subjected. The Cross Timbers extend unbroken from the upper course of the Colorado to the Red Fork of the Arkansas, since they cross the Trinity at its spring above the mouth of the False Washita.

SECTION III
PRODUCTS OF THE PLANT & ANIMAL KINGDOM

Generally speaking, the natural products of the soil of Texas are the same as those of the adjoining Southwestern States of the Union, particularly Louisiana. Cotton and sugar are the chief articles of export. Both are of excellent quality and Texas cotton is in great demand. It is cultivated primarily in the fertile valleys of the Trinity, the Brazos, the Colorado and the Guadalupe, and also a great distance from the coast in the undulating regions on the Brazos, for example, near its falls and on the Colorado near Austin. The quantity of cotton harvested in 1847 is estimated at ten million pounds. The cultivation of sugarcane is confined to a few strips of the richest alluvial soil in the lower coastal country, particularly in the bottom of the Brazos and of a few smaller rivers, such as the Caney, the St. Bernard, etc.

In addition to cotton and sugar, tobacco is also exported, but so far only in small quantities. The greater part of Texas is suited to the cultivation of tobacco and since it is of excellent quality, it will be an important staple article in years to come.

Among the products which are used for home consumption, corn plays the most important role. It forms the chief article of food for man and beast. That used for human consumption is ground by handmills or in some other manner and, in accord with the usual custom in the United States, it is baked into cornbread which is usually eaten warm; or the kernels are scalded until the skin readily peels off, and after it is removed they are cooked into a kind of mush called hominy. Horses are fed corn throughout North America instead of oats, and oxen also receive some, in addition to hay or grass.

The large hybrid variety of corn with ears a foot in length and with large four-cornered kernels, flat or retuse on the anterior side, and wedge-shaped at the posterior ends, which is also cultivated primarily in the Western and Southwestern States, is cultivated here. The yield of corn is extraordinarily large in Texas and surpasses by far all the European varieties of grain.

The latter can also be raised in Texas. Oats are grown successfully in the upper valley of the Colorado near Bastrop. According to reports, wheat is raised extensively on the Trinity near its forks and thrives well. However, the cultivation of these two kinds of grain will scarcely reach any magnitude in Texas, since their cultivation in comparison to corn involves so much more work. It will also be difficult to compete with the western wheat-growing states, where wages are lower than in Texas. At the present time, mills for grinding wheat and rye are lacking.

Most of the vegetables found in Germany also thrive here and some, such as beans and peas, grow luxuriantly. However, the climate does not seem to be suited for raising potatoes, for they have an insipid taste. This also seems to be the case in other Southern States of the Union and only in the far Northern States such as Maine, are good potatoes raised. They are replaced by the sweet potato (*Convolvulus batatas L.*) which is also raised in other parts of the United States as far as New Jersey, but which finds exceptionally adaptable soil in Texas. The yield is often extraordinarily large and surpasses by far that of the potato, which it resembles in shape and taste, although the plant, as is known, belongs to a different family. Among the fruits raised in the gardens particularly adapted to the climate are watermelons (*Cucumis citrullus L.*) and various kinds of edible pumpkins which are grown quite extensively. The climate is too warm for most of the fruit trees raised in Germany, especially apples and pears. Only the peach thrives well here, as in most other parts of the United States, although the fruit is usually small and unimproved, owing to the neglect which the careless American farmer accords his gardens and fruit trees. In some years the peaches are so plentiful that the settlers feed them to the swine.

Fig trees are found everywhere in Texas and grow into quite large trees, but they are nowhere planted in large numbers. For some reason orange trees, while they thrive well in adjoining Louisiana and bear ripe fruit, do not seem to do well out of doors in Texas. At any rate, orange groves planted at several places, as, for instance, on Galveston Bay, did not prove a success. Perhaps the peculiar cold northers of the Texas winters are the cause of this. On the other hand, not enough trials have been made to introduce them to justify a definite conclusion.

Texas is not rich in native, edible fruits. Strawberries are unknown. Various kinds of grapes are found everywhere and climb up the highest trees. The most common kind (*Vitis labrusca L.*) bears large blue grapes similar to the most beautiful Burgundy grapes, but the thick skin is very astringent and no palatable wine can be made of it, although no doubt a good, useful vinegar could be produced.

That wild grapes are so plentiful is no proof that the land is suitable for the cultivation of tame grapes. Wild grapes are also very common in many parts of the United States, especially in the West, but the cultivation of the improved varieties has so far proven successful in but few places, as, for instance, near Cincinnati. Rather would the fact mentioned by Baron Alexander von Humboldt, that grapes which produce good wine were grown for a long time at Paso del Norte on the Rio Grande, augur well for direct trials in the adjoining Texas.

Everywhere in Texas, especially in river bottoms, grow pecan trees (*Carya olivaeformis Nutt.*) which bear fruit resembling walnuts. Whole wagon loads of these are gathered in the fall and brought to Houston, from which point they are shipped to the Northern States.

The products of the animal kingdom are also very numerous. Texas is a natural grazing land, unsurpassed by any in the world. The cattle thrive and multiply in a remarkable degree in the wide prairies of the level coastal country, as well as in the undulating hilly country and without the care and help of human beings. They find nourishment in the open throughout the year and barns for them are unknown. If carried on on a large scale, Texas could furnish the entire United States with the necessary meat supply. The country is also well adapted for raising horses. This is demonstrated by the fact that horses which the first Spanish settlers brought into this country reverted again to such a wild state that their descendants even to this day are now living here in a wild state with buffalo in the uninhabited prairies.

Hogs, which run around here with perfect freedom, as is also the case in other parts of the United States, multiply quickly, and since their flesh is the chief food of the English-American settler, they are raised in large numbers. However, the forested Western States offer more opportunities for breeding them than does the greater part of Texas with its open prairies.

So far few sheep are found in Texas. However, the several attempts that have been made and the fact that in the neighboring provinces of Mexico, particularly along the Rio Grande, they are raised on a large scale, establishes beyond a doubt, that also in Texas, at any rate on the dry prairies of the undulating highland and the higher plateau, sheep can be successfully raised.

Bee culture, which already receives much attention in certain sections of Texas, could become a very important industry of the country if engaged in generally, since the wealth of flowers in the extensive prairies invites the keeping of bees.

SECTION IV
PRODUCTS OF THE MINERAL KINGDOM

The accessible and settled portions of Texas are poor in valuable minerals. This does not seem strange when considering the geological structure of the country, since the level coast country is a product of recent alluvial deposits, while the undulating hilly country is composed of rocks of the fletz formation. A bed of brown coal is supposed to exist on the upper course of the Trinity. Genuine coal has so far not been found in Texas. However, during the past few years, a rich bed of excellent coal is supposed to have been discovered on the other side of the Rio Grande near Revilla or Guerrero and it is contemplated to use it on steamboats plying on the Rio Grande.

What truth there is to the report of a wealth of precious minerals in the higher hilly country, particularly the fabled silver mines in the region of the San Saba River, will be revealed by the more detailed information to be found in the later pages of this book. It will in no way confirm any such conjecture.

Even without a wealth of minerals, Texas is a highly favored land with its unsurpassed fertile soil and its mild climate.

SECTION V
INHABITANTS

The settlements of the Spaniards extended over a small portion of Texas and were confined in reality only to a few settlements in and near San Antonio de Bexar, as well as at several unimportant points on the coast, such as Goliad, Refugio, etc., and Nacogdoches near the Red River, which was primarily a military post. The entire population of Texas under Spanish rule, numbered only about six thousand souls according to Scherpf's computation. Since then the Spanish and Mexican population has not increased, but on the contrary has decreased continually after the influx of the Anglo-American settlers from the United States. Nearly all the well-to-do Mexican families have recrossed the Rio Grande in order not to be near the arch enemies of their nation, the "Americanos del Norte." At the present time the Mexican population in Texas proper may not number more than two thousand souls, of which about a thousand live in and near San Antonio. Among these are only a few wealthy and cultured families. Many of them belong to the lower class and their color and features betray plainly, that in addition to their Castilian blood, very much Indian blood courses in their veins. This mixed race has the faults as well as the good traits of the Mexican people; however, the former predominate. They are lazy, revengeful, cruel and overbearing toward the conquered foe, but cowardly and cringing toward the victorious one. Among their good qualities belong their temperance in the use of alcoholic drinks and their polite obligingness toward the stranger.

The Anglo-American settlers form by far the greater part of the population of Texas. Since the year 1819, when the enterprising Moses Austin from the Northern States carried out his agreement with the Republic of Mexico to bring in colonists, the immigration to Texas has continued uninterrupted.

Up to the year 1830, the number of settlers, according to Scherpf, was about 15,000. They lived on an area which, when compared with the average European population, could easily have supported fifteen million people. According to the state census of 1847, the population of Texas was 143,205 souls of which 39,060 were negro slaves. Most of the emigrated Americans came from the Southern and Southwestern States, Louisiana, Alabama, Arkansas, Missouri, etc. Emigrants from other States, east of the Hudson are called Yankees and are found in the larger and smaller cities engaged as merchants and manufacturers.

Habits, customs and speech throughout the entire State are like the Anglo-Americans, especially those of the Southwestern States, with the exception of several regions where settlers of foreign nationalities live to themselves. The government of the State is based upon the constitution of the State of Texas adopted in convention at the city of Austin, 1845, and which was later ratified by the Congress of the United States. In a general way it contains all the chief provisions of the Democratic constitution of other States of the Union, particularly those of the younger western ones. Texas is a slaveholding state, and all peculiarities of the private and public mode of living which this institution —as it is euphemistically called in the constitution of the slave states—brings with it in other Southwestern States of the Union, is also found in Texas. In addition to slaveowning planters, there are also a larger number of farmers who till their own soil.

While the original American population of Texas without doubt was composed greatly of the most degraded riff-raff, adventurers, gamblers, swindlers and murderers—the scum not only of the United States but of all nations—present morals and respect for the laws of the land are as a general rule not any lower than in the adjoining Southwestern States. In comparison to the Northern and Eastern States, deeds of violence, that is, crimes against life are more common, owing to the passionate nature of the people of the South. With the exception of premeditated murder, these crimes are less punishable in the eyes of the people.

In the extreme border settlements the law-enforcing agencies are often powerless and in those places occasional lynchings, known as lynch laws (popular justice or self-help) occur even to this day.

SECTION VI
ABORIGINAL INHABITANTS, OR, INDIANS

In addition to people of European origin, aborigines or Indians also inhabit Texas. These belong to several tribes which distinguish themselves upon closer examination according to their peculiarities, mode of living and custom, but all have in common the unmistakable features of the Red Race.

None of the various tribes in Texas are as tractable and docile as the Indians in the interior of Mexico; on the contrary, all share the warlike and independent inclinations so characteristic of the North American Indian. Nevertheless, there are degrees of savageness among the tribes, and the Texas settler distinguishes between "tame and wild Indians." The former are all such who acknowledge the authority of the Texas government in silence and submit to a certain amount of supervision. They have permission to travel about within the boundaries of the settlements and to hunt. Several of these tribes have an officer of the United States Government, called an Indian Agent, who is constantly with them, who accompanies them on all their journeys, and is responsible for their peaceful behavior toward the settlers. The wild Indian tribes, however, take a decided hostile attitude toward the whites, or, if they have entered into an agreement with them, their agreement is not guarantee enough for their peaceful behavior so as to allow them to travel about in the settled portions of the State. To the former belong the Lepans or Lipans, who are related to the Comanches in language and customs, and the Tonkohuas, both small tribes, who number only a few hundred souls. To these semi-civilized tribes must also be added the Caddoes, who live in a village on the upper Brazos. The Shawnees cannot be classified among the real Texas Indians, since they live farther north and travel through the country only in small bands on hunting expeditions. They also belong to those who are partially civilized.

Among the wild tribes, the Comanches are by far the most
dangerous on account of their numbers and warlike disposition. They are probably the most virile of all American
tribes. In the later pages of this book they will be described
more in detail. Less numerous are the Wacoes, who number
several hundred souls. They live in open warfare with the
white settlers and their depredations and thefts are carried
out with such slyness that their participation in these raids
can seldom be definitely established.

The Wacoes were accused of nearly all the depredations
and acts of violence perpetrated during my stay in Texas.
The murders of two former officers, Captain von Wrede
and Lieut. Claren, between Austin and New Braunfels were
charged to the Wacoes.

The Karankuhas or Carankoways, a ferocious tribe, which
traveled along the coast ten years prior to this time, seem to
have vanished. It is reported that they had destroyed one another in their wars among themselves. It is told that they ate
the flesh of their enemies, and a trustworthy friend assured
me that he had witnessed such an act. The cannibalistic prac-
tices were also attributed to the Towakhanies, a small tribe.

The total number of Indians living in Texas is not more
than 8,000 to 10,000, half of which are Comanches. An
accurate count of the Indians is impossible and only an
approximate estimate can be made by such people who as-
sociate with the various tribes, such as the Indian Agents,
who are appointed by the government of the United States
and whose duty it is to supervise and keep them in check.

It is apparent that Indians of Texas, as well as the Red
Race in North America, will be driven from their homes
and eventually be exterminated. The most beautiful lands
have already been taken from them, and restlessly moving
forward, the white, greedy conqueror is stretching forth his
hand to take their new hunting ground also. Several tribes
have already vanished within the memory of man and other
tribes, once numerous and powerful, have shrunk to unimportant remnants. The Comanches very likely will hold

out longest in their stony plateau. The prevailing infertility of the soil and the inaccessibility of their homes will serve to protect them more against extermination than will their number and warlike disposition.

SECTION VII
HISTORY OF THE GERMAN COLONIES IN TEXAS & IN PARTICULAR THOSE ESTABLISHED BY THE MANZER VEREIN

In the very beginning when all Texas was made accessible to foreign colonization, Germans came to Texas with the North American colonists and settled separately in various parts of the State. Several Germans took part in the War of Texas Independence against Mexico and one of them became chronicler of these remarkable battles.

Many German families live in the central part of the State, Some had lived there for the past ten or twelve years between the Brazos and the Colorado, and particularly near the hamlets Industry and Cummins Creek, and on both sides of the road leading from San Felipe de Austin toward La Grange. Most of them cultivate their own fields or do so with the help of white labor. Very few are plantation owners and possess negro slaves.

Although each year brought new immigrants from Germany, their number was always comparatively small, until in the spring of 1844, when an organization was formed which had as its objective a systematized immigration from Germany to Texas on a larger scale. This organization consisted of German princes, barons and noblemen, who in the spring of 1844 organized in Mayence under the name Society for the Protection of German Immigrants in Texas.

The purpose of this society was stated in the program of April 9, 1844, as follows: "To direct German immigration as much as possible to one favorable location; to offer the immigrants assistance on their long journey and in their new home; and to work energetically to the end that they may be assured a new home across the sea."

The same program mentions further, "that after careful consideration, the Verein had decided that Texas was best suited to the German immigrants." Experienced men, such as Prince von Leiningen and Count von Boos-Waldeck, sent out by the society, had traveled through Texas and their complete report thoroughly convinced the Verein, that they could make this selection with a clear conscience.

The Verein guaranteed each adult male immigrant an outright gift of one hundred sixty acres of land and every family three hundred twenty acres. In another document, issued the same year by the society and entitled "Conditions for the Immigrants," such a number of advantageous promises and glowing reports concerning the tract of land to which the immigrants were to be directed were made, that if the former could have been fulfilled, and presupposing the correctness of the latter, every immigrant who joined the Verein could regard his fortune as already made. It was therefore an easy matter to find one hundred fifty families who were to be accepted as the first contingent. They arrived safely in Galveston and were received by Prince Carl zu Solms-Braunfels, who had been given the supreme command of the colonization project. In the meantime a vital obstacle had already arisen in the carrying out of the original plan. The land to be colonized, which was to contain four hundred fifty square miles in a fertile region of West Texas, had been bought from a Frenchman, a certain Bourgeois d'Orvanne who, in addition to the sales price, was guaranteed a voice in the management of the colony. Immediately after the first contingent had arrived in Texas, it was discovered that the supposed owner of this tract of land, as large as a kingdom, was nothing but a frivolous adventurer who did not own a single acre of ground. He had actually received from the Texas government a so-called grant, situated northwest from San Antonio de Bexar, with the proviso that he colonize it within a specified time. But all claim to this grant had long since been voided, since the terms of the contract had not been met.

The Verein thus found itself suddenly in the predicament of having no land to which to direct the immigrants, although a considerable number had already arrived in Texas through its efforts. A second contract was signed June, 1844, with a German who had lived in Texas for a number of years, Mr. Henry Fischer of Houston. This contract pertained to a certain tract of land north of the Llano River on the right bank of the Colorado which Mr. Fischer and another German by the name of Mueller had also received from the government of Texas. The Verein paid a rather high price and, in addition to this, the vendors were guaranteed certain rights in the development of the colony, as well as a share of the profits expected from the undertaking. However, this did not remove all the difficulties for the first settlement. It soon became evident that the tract of land just purchased was too far removed from the coast and from the inhabited cultivated region that it was not deemed advisable to take the immigrants there immediately. This immediate difficulty was solved by Prince Solms who bought a tract of land containing several thousand acres on the road from San Antonio de Bexar to Austin, which was to serve, as it were, as a halting place until the colonizing of the actual Verein territory could proceed.

The selection of this place, upon which the hamlet New Braunfels has since been erected, must be regarded as a most fortunate one and appears as a bright spot among the blunders and mistakes found in the history of the German colonization in Texas by the Mainzer Verein. Through it at least a small portion of the extensive colonization plan has materialized, since a colony of Germans is founded which can maintain itself and which to all appearances will form the nucleus for German settlements round about. A more suitable location could not have been found in all of Western Texas according to my opinion as well as that of many

Americans. May the much abused Prince Solms at least receive due credit for this act.

To every one of the immigrants of the first contingent was
allotted a half-acre tract for a home site in the proposed town,
also ten acres of land in the vicinity of the town and this without detriment to his claim to the one hundred sixty and three
hundred twenty acres, respectively, promised in the grant.
The people, glad to have found a permanent place at last after wandering about in the wilderness for so long a time, began with the erecting and arranging of their homes in good
spirits. In the meantime difficulties of another nature arose,
increasing as time went on and becoming the direct and immediate cause for the failure of the entire venture. The funds
from the Verein began to flow more sparingly, thereby hin-
dering and embarrassing the management in Texas.

A capital of several hundred thousand gulden raised by
members of the Verein through subscriptions appeared as
a large sum to the promoters in Germany. When, however,
one considers that this amount had to take care of hundreds
of people for months and even years in a land across the sea
where many necessities cost double and treble as much as
in most European countries, of transportation into the interior of a wild and almost uninhabited country where every
communication with the coast meant added expense, and
of all further requirements incident to the founding of a
new home—then even such an amount appears small and
is quickly used up.

Prince Solms, no doubt foreseeing the difficulties which
a continuation of the venture presented, resigned. Baron
Otfried von Meusebach of Potsdam, formerly Prussian
governmental assessor, was appointed as his successor. He
came to Texas during the summer of 1845 and took charge
of the Verein affairs. Prince Solms returned to Europe. That
he unselfishly sought to discharge the duties imposed on
him and that he did so at great personal sacrifice, even his
enemies had to admit. However, he lacked that business
ability so necessary for such a venture, for which lack the

good intention to serve his German compatriots to the best of his ability did not make amends. At all events he would not have been equal to the occasion of handling the difficulties and straitened circumstances, which became especially acute after his departure, nor would he have been as able as his resourceful successor to devise new ways and means to meet them.

His chivalrous, romantic nature at which his enemies cavilled does not, in my opinion, deserve severe censure, for though his general demeanor may have appeared singular and quaint to those who did not share similar views and disposition, it was quite harmless and in no way could hurt the affairs of the Verein.

Rather culpable was his decided antipathy against and disdain for the American national character, which the Prince seldom concealed even in contact with Americans. Since the American was already more or less jealous of the foreigner, it is easy to see how such an attitude would engender a hostile feeling toward the entire Verein program and to what extent such hostility could prove detrimental and ruinous is not hard to see when one considers that the arrivals were in so many ways dependent on the native resident.

Von Meusebach began by introducing business standards and a more careful system of bookkeeping, as well as by curtailing the services rendered the immigrants to the most essential. The firm belief that the success of the venture could be hoped for only if the strictest economy were observed with the funds, hardly adequate as they were for the carrying out of the magnificent plan, justified these rigid measures. These procedures, nevertheless, caused considerable dissatisfaction and made von Meusebach's administration unpopular from the very beginning. Of course, one must not forget that in view of the contract, which definitely established their claim, the settlers had some ground for dissatisfaction, since the Verein had not fulfilled its obligation and promises to them. It was particularly unable to assign to each family the three hundred twenty acres suitable for im-

mediate cultivation. The demands of the colonists for continued support through the Verein until the promise of the
three hundred twenty acres had been fulfilled were, therefore, seemingly well-founded. But in an undertaking such
as that of the Verein, the many difficulties could not have
been foreseen in Europe, hence fairness as well as mutual
interests demanded that neither side insist too stubbornly
on its rights but that, in view of the circumstances, each side
make concessions. Unfortunately, however, such a mutual
arrangement did not ensue generally and the strained relations between many of the immigrants, disappointed in
their expectations, and the management grew more pronounced.

In the fall of the same year von Meusebach made an expedition of several weeks' duration to the Indian territory
north of New Braunfels for the purpose of selecting a place
for another settlement. The place chosen was north of the
Pedernales (a tributary of the Colorado) about ninety miles,
or a three days' journey on horseback, distant from New
Braunfels. At this point the headrights (i.e. land grants of
a definite number of acres which the State of Texas gave to
individuals for services rendered during the war,) previously
purchased, were located. Thus a tract of several thousand
acres was acquired for the Verein. Here, later the village of
Fredericksburg was built.

On his return from this expedition, von Meusebach
found the news awaiting him at New Braunfels that several thousand immigrants would arrive very shortly at
Galveston and that some had already arrived. He departed
at once in order to make the necessary arrangements to
accommodate them temporarily and to arrange for their
further transportation. This, however, was a difficult task
for, inconceivable as it is, while the management in Germany had indeed sent over thousands of people, it sent no
money for their maintenance nor for their transportation
to the settlement. Nor could this have been a mere over-
sight for not even those monies were sent which many of

the immigrants had deposited with the understanding that they would be refunded on demand upon arrival in Texas.

This latter circumstance was the cause of much embarrassment to some families who, upon their arrival at Galveston, found themselves without any funds, although they had deposited several thousand dollars with the Verein in Germany. Many also lost months of valuable time until, after many futile efforts, their property was finally returned to them. This action of the directors in Germany in withholding trust funds through misrepresentation from the rightful owners, to their great injury, deserves the most severe censure. That it caused a storm of protest from those affected was but natural.

In the meantime the newly arrived immigrants were gradually brought from Galveston to Indian Point (Indianola,) a harbor on Lavaca Bay, from which point later they were to be sent to their final destination.

The funds of the Verein were just sufficient to supply the immigrants at Indian Point with victuals for a few months. But that was all that could be done for the time being. The necessary funds were lacking to transport them into the interior, but even if this had not been the case it would have been inadvisable to bring thousands of people many days' journey inland without having the assurance that funds would be available to maintain them. It was much easier and less expensive to keep them on the coast. In the meantime the spring of 1846 arrived and with it the heat of a semitropical climate. About three thousand of the poor immigrants lay crowded on the sandy coast, without an adequate water supply and fuel, living in sod houses or tents which afforded no protection against the rain nor against the hot rays of the sun. The long idleness and the uncertainty as to the future had a demoralizing effect upon them. All conditions combined there for climatic sicknesses to gain a foothold. These did not tarry in seizing their victims. Malaria, bilious fever and dysentery soon became general and the mortality increased with alarming rapidity. A general cry of distress arose to leave this dreadful place. But this became more difficult now than before. War with Mexico, having been declared in the meantime, meant that all available transports for the moving of war supplies and provisions were requisitioned. Even if conveyances had been there in sufficient numbers, they could not have been engaged, since money was lacking and the credit of the Verein had reached its limit.

A contract which von Meusebach had made with merchants of Houston, the Torrey brothers, to move all the immigrants from the coast could not be carried out entirely, since a continual rain, very unusual during this time of the year, had made the roads in the low coastal country impassable and detained the wagons, drawn by oxen, often for weeks at the river crossings. In their despair several hundred immigrants formed a volunteer corps and fought against Mexico. The remainder now started on their way to New Braunfels singly or in family groups, utilizing the few wagons which gradually arrived but abandoning their property which they had guarded so carefully up to this time.

On the journey from Indian Point to New Braunfels, which required several weeks, due to the slow-moving wagons drawn by oxen, various epidemics and sicknesses took their heaviest toll. Entire families died during the journey from the coast to New Braunfels, and the course along the Guadalupe was marked by countless German graves. All moral ties were dissolved and the prairie was witness to deeds of violence from which the natural feelings revolt and which sullied the German name in the far away country. Many died along the way; others after they had dragged their emaciated bodies as far as New Braunfels. A large number were scattered in the interior of the country among the American settlers and were not seen again. Still others, whose means permitted it, returned brokenhearted and discouraged to Germany. It is certain that in the few summer months of the year 1846 more than one thousand out of four thousand German immigrants, who had come to Texas in the fall of 1845 under the protection of the Mainzer Verein, died and not more than one thousand two hundred actually settled upon the land secured by the Verein. German immigration has not been able to recover from this terrible blow and the number of immigrants from Germany has since been very small.

In the course of the same summer, however, the undertaking of the Verein had been more successful in other sections. New Braunfels had grown rapidly despite all difficulties and gave promise of becoming a thriving city. Fredericksburg was founded on the Pedernales and grew despite even greater difficulties, numbering close to a thousand inhabitants by winter. All this was accomplished in the face of a continued shortage of money, which was only occasionally relieved by a remittance from Germany to satisfy the most pressing creditors.

In January and February of the year 1847, the long-planned expedition into the original Verein territory or grant north of the Llano was undertaken, since all reliable information concerning it was lacking. Through this expedition I, as well as others, became convinced that by far the greater portion of this land was infertile and not at all suited for farming. This condition, coupled with the fact that it is so far distant from the coast and the cultivated sections of the country, convinced us the more that its selection as a site for a German colonization on a large scale was ill chosen. I left Texas shortly after this and have no knowledge of the further developments of the Verein undertaking through personal observations, but only from letters and verbal reports. Herr von Meusebach resigned as commissioner-general, since he had long harbored the wish to do so, especially since the time when the population of New Braunfels had held a hostile demonstration against him. Not long after this the Mainzer Verein dismissed its officers and agents in Texas and its real estate and grants became property of another company, The German Colonization Society for Texas at Biebrich. The latter, without the financial backing which the Mainzer Verein had, will be even less in a position to carry out the grand, but from the beginning mismanaged, undertaking.

Thus was the extensive plan of the Mainzer Verein for the protection of the German immigrants in Texas, which was at first hailed with delight by many Germans, abandoned as a failure. The number of unsuccessful German colonization projects is now augmented by another.

SECTION VIII
CAUSES FOR THE FAILURE OF THE MANZER VEREIN UNDERTAKING

It seems proper in the interest of future undertakings of similar nature, to inquire as to the causes of its tragic ending. I believe the following reasons may be advanced as the chief causes. The greatest mistake which could not be rectified later was the fact that no thorough and authentic information of the strip of land to be colonized had been secured.

It is hardly believable but true: the Mainzer Verein sent thousands of people across the sea to settle on a strip of land which none of its members nor any of its agents had ever seen. In fact, not even the persons from whom it had been bought had approached it within several days' journey. It was their duty to obtain accurate information of the natural conditions of the land upon which the settlement was to be made before sending over the first settlers. If that had been done, they would not have tried to colonize land several days' journey from the nearest settlements, inhabited by marauding, dangerous Indians and, finally, which was for the greater part rocky and sterile, and less suited to agriculture than other regions having more direct communication with the coast.

If the natural, as well as other, conditions of the country to be colonized had been more accurately known, the Verein would have been more careful in its promises made to the colonists. It would then have been apparent that the fulfillment of such promises, contained in the program of the Verein, would in part have been impossible of fulfillment or would have required too great a sacrifice to carry out. For example, it would have been known that the transportation, by the Verein, of the colonists from the coast to their destination could not have been carried out for the stipulated price of $8.00 for each person, without some limit on the weight of the baggage. This not having been done, the Verein suffered a considerable pecuniary loss, since many carried with them worthless and useless baggage which more than trebled the cost of transportation.

Another mistake was the fact that the plan was too elaborate when considering the funds on hand. Obviously, the estimate of the probable cost was entirely too low in the original plan. When these estimates had to be exceeded continually while carrying out the plan, the funds were soon inadequate. Extraordinary measures had to be employed to relieve the situation. Before these could be carried out in Germany, weeks and months passed and during this time the undertaking suffered irreparable losses. In consequence of this the credit of the Verein in Texas was impaired and all operations of the management were hampered. The inadequacy of the funds was also caused by the unsuitable location selected for a settlement. Huge expenditures were also incurred by the Verein in that the colonists could not be transported immediately on their arrival in Texas to the promised land. As a result of this the Verein had to assume the cost of providing for the immigrants, some of them for more than a year.

The funds at that time were altogether out of proportion to the magnitude of the object to be accomplished, when, in the fall of 1845, several thousand immigrants were sent to Texas in one contingent and no adequate sum was put at the disposal of the management in Texas to transport and maintain such a large number of people. The Verein had not even made provision to refund the monies which the immigrants had deposited in Germany.

Finally, several circumstances which the directors of the Verein could neither foresee nor prevent also contributed to the failure of the entire undertaking has already been mentioned. The outbreak of the war between the United States and Mexico, as well as the unusually wet spring of 1846, prevented the timely transportation of the immigrants from the low coastal country to the more healthful higher region, and thus led to the sad catastrophe which gave the German immigrations to Texas in general, and those under the protection of the Mainzer Verein in particular, a severe jolt.

SECTION IX
ADVANTAGES & DISADVANTAGES OF GERMAN IMMIGRATION TO TEXAS

The failure of the extensive colonization undertaking of the Mainzer Verein can moreover not be used as a criterion in answering the question whether immigration to Texas should be at all encouraged. It would be better to view the advantages and disadvantages entirely separate from the outcome of that undertaking.

Among the advantages and attractions which would argue in favor of settling in Texas, I would like to place above all the mild and excellent climate. The climate of Texas is surely one of the best in the world. During the greater part of the year the sky is clear and blue. A real winter is unknown and only during a few months of the year does the temperature at times go below the freezing point. Whereas it is in itself a great advantage to live under such a clear sky during the winter months, the mildness of the climate offers also other advantages and relief to the European colonist. Shelter and clothing are more easily obtained than in the rough climate with a severe winter. The houses are usually built of light wood, since they serve more to protect against the rays of the sun and rain than to protect against the cold of the winter. Small houses are also adequate, since much work can be done out of doors which would require a separate room in the northern regions. The settler needs no shelter for the cattle, for they can remain out of doors throughout the year without any great discomfort. Only for the horses is a lightly constructed barn built. Neither is it necessary, as is the case in the northern climate, to gather great supplies of feed for cattle, for they can find ample food out of doors even during the winter months.

The extraordinary advantages which are offered the settlers in cattle raising cannot be estimated too highly. The prairies are the most natural luxuriant pasturage which any country could offer and owing to the sparse population, each settler has access to as much grazing land as he desires. The cattle thrive excellently and are not subject to sicknesses. Now and then a head is lost when maggots of the fly penetrate a wound, accidently incurred, to such an extent as to reach a vital organ. It it not necessary to care for the cattle. They care for themselves year-in, year-out, roaming about, often wandering many miles from the home of the owner.

The natural conditions for agriculture are scarcely less favorable than cattle raising. The open, unforested land spares the settlers the difficult task of clearing a field which is the first requisite in nearly all parts of the Northern States of the Union. The settler in Texas, after he has fenced the field chosen for cultivation, need only plow the sod with a heavy plow drawn by two or three yoke of oxen. He can then plant corn and other seeds without further preparation.

In several prairies, called weed prairies which are not covered with a mat of grass but only with certain perennial plants, it is necessary to loosen the topsoil only after the weeds have been burned off, in order to plant the seeds.

Fertilizing the fields in Texas is unnecessary. The natural fertility of the soil in most parts of the State assures excellent crops for years to come. A failure in the chief crops, such as corn, is unknown. Occasionally there is a partial crop failure in cotton and sugarcane, which is cultivated by negro labor.

The present cheap price at which farm land may still be purchased by the immigrants also favors the colonization by European settlers. Land in Western Texas, even near the settled regions, can be bought for from two to five dollars per acre, and in the more remote regions the price is still lower. It even happens that owners of large tracts of land offer the immigrants land free of charge in order to enhance the value of other land.

Finally, I would like to mention that due to the sparse population, it is easy for immigrants of one nation to live as a unit in a community, thus having the advantage of living among their compatriots, and retaining the customs and habits of their native land.

On the other hand, I do not wish to keep silent as to the disadvantages and difficulties which meet the European settler, particularly the German. The climate is so different from that of northern Europe, that at first it reacts rather severely upon the German constitution. Of all the German immigrants who came to Texas during my stay, I can scarcely name one whose health was not affected by the change of climate, even if only manifesting itself in a more or less languid feeling, loss of strength and noticeable loss of weight.

I have often heard German laborers complain that while they were not indisposed, they nevertheless did not feel as vigorous and inclined to work as in Germany. As a rule, one can readily distinguish the German who has been in Texas a longer period of time from the newly arrived immigrant by the paler face and leaner body.

Frequently the acclimatization is not so gradual and imperceptible, but more or less violent and dangerous symptoms appear. Bilious and malaria fever prefer the newly arrived European, although the natives are also not spared. The strong and healthy in particular contract a virulent form of disease.

When the acclimatization has been accomplished in a mild or virulent form and the person, used to the northern climate has accustomed himself to the conditions of the warmer zone, the greater danger has passed, but he is not immune from other sicknesses.

Among the prevalent diseases, malaria and ague take first rank here as is the case more or less in all Southern and Western States of the Union of North America. Although seldom directly fatal, their repeated recurrence becomes dangerous and undermines the strongest constitution. They are prevalent along the coast and the lower forested river bottoms, but also found in the higher sections of the country where the conditions, attributed as their cause, are not at all present. All bilious fevers, however, which occur during the hot summer months in all sections of the country, are extremely dangerous. The course of this disease runs exceptionally fast and requires immediate and energetic attention, otherwise it often proves fatal within twenty-four to forty-eight hours.

Dysentery also seems to be distributed widely and occurs during the fall months in a very malignant form. However, according to the statements of the older settlers, it has never been as general and in an epidemic form as in the summer of 1846 in Western Texas where many German settlers were snatched away by it. The cause for this may be attributed to unusual and peculiar conditions.

It should also be stressed, on the other hand, that many diseases prevalent in Northern Europe are wholly unknown in Texas. The common and often fatal typhus fever occurs seldom, and equally seldom are lung trouble and consumption in particular, which claim so many victims at home. As a matter of fact, many persons who are afflicted with lung trouble feel perceptibly better and recover completely when coming from a northern climate to Texas.

Nevertheless, Texas cannot be regarded, generally speaking, as healthful as Northern Europe and particularly Germany. Every German—despite the lovely alluring climate—assumes a considerable risk when immigrating to Texas. At any rate the danger is greater than for the immigrant to the more Northern and Northwestern States of the Union, as for example Wisconsin, Illinois,

Missouri, etc. Even if the climate in these States, which is similar to that of the native land, does not agree with the German constitution, malaria and other sickness to which the newcomer is subjected do not occur in such a virulent and dangerous form as in Texas. The acclimatization also proceeds less noticeably and without danger, due to the similarity in climate. I do not wish to assert that the climate of Texas is harmful to every German. There are many German settlers living in the central part of Texas between the Brazos and the Colorado who for the past ten years have enjoyed good health. However, no one has kept a record of the great number who came over with them and who have died of sickness.

It is also certain that the injurious influence of the Texas climate can be prevented to a great extent in the very beginning by a careful and regulated mode of living. A moderate diet, and abstinence from all alcoholic drinks as well as exercising care not to subject one's self to the violent rays of the sun during the hottest part of the day are some of the precautions to be recommended.

The climate of Texas, finally, is no more injurious to a German than the climate of neighboring Southern States, particularly Louisiana, Mississippi and Alabama. On the contrary, when compared with Louisiana, Texas would probably receive preference. Such broad, low river bottoms and extensive swamps as are found in Louisiana, which are the source of mephitical evaporations, are not found in Texas. This is especially true in Western Texas where swampy regions and stagnant water are unknown.

Another disadvantage of settling in Texas, particularly in comparison with Northwestern States, is the imperfect communication between the coast and the interior. Texas, as mentioned previously, has few natural water courses suited for commerce when compared with other States. The Trinity is the only river navigable for a long distance when not taking into account the Red River and the Rio Grande which is navigable in its lower course only.

Texas possesses neither highways nor railroads and all communication is carried on over roads, whose beginning was as simple as their upkeep is now. In reality, they do nothing more than indicate the direction through the wilderness by the wagon and hoof tracks of those who have gone before. In the same degree to which the Western States in the basin of the Mississippi owe their rapid growth to a great extent to the excellent natural water communications, the development and use of the extraordinary natural resources of Texas will be hampered and hindered for lack of them. Every settler will suffer on account of this disadvantage. This is felt more keenly because the more healthful and beautiful places of the State, which invite the settler, are so far removed from the coast.

That Texas is a slaveholding State may also be considered as a disadvantage when comparing it with the free Western and Northwestern States of North America. In all slaveholding States, the social standing of the settler who is not a slave owner, is lower than in those of the free States. The labor of the free man is not respected and he takes a subordinate position when compared to the slaveholding planter. Aside from the fact that there is a natural aversion to owning slaves, comparatively few German immigrants will be able to buy slaves. The great majority will have to perform their own work and will, therefore, suffer from this social disadvantage.

Settling in a slave State appears not to offer any advantages, since experiences have taught that the free States grow more rapidly in wealth and population, and in consequence of this the value of property of the individual increases rapidly.

When weighing the advantages and disadvantages of settling in Texas, especially as compared to settling in the more northern States, no doubt the advantages in Texas outweigh the others when not taking the danger to health into account. If one, however, places a greater value upon health and, therefore, prefers to live in a more healthful climate, where he can also establish himself, but with greater efforts, he had better select the Western or Northern States, particularly Wisconsin, Iowa, Illinois and Missouri.

SECTION X
WRITINGS ABOUT TEXAS

Although this country has attracted general attention for only a comparatively short time, there are already a number of books on Texas available. Very few, however, are of lasting scientific value. In most instances the chief aim was not to acquaint the reader with the natural conditions of the country, but they were to serve the public only as light reading. The authors, therefore, mixed fiction with facts as they saw fit to arouse the interest of the reader. Often the author was only concerned to arouse a favorable or unfavorable opinion about the country, irrespective of the truth, in order to serve some selfish purpose. To the latter class belonged various "leaders and advisors" to Texas immigrants of whom quite a number appeared in 1846. The crude untruths and fabulous exaggerations found in the majority of these books, would arouse the mirth of those who have seen these things with their own eyes were it not for the fact that the unscrupulous methods of the authors who have caused the poor immigrants many bitter disappointments through misrepresentations, did not call forth a feeling of resentment and indignation.

To this type of literature belongs a book, which due to the literary renown of its author has been read extensively in England, as well as in Germany. It is entitled *Narrative of the Travels and Adventures of Monsieur Violet in California, Sonora, and Western Texas* by Capt. Marryat. Fiction and facts are so interwoven that they cannot be distinguished from each other. The fiction, which deals with the wonderful adventures and fabulous exaggerations are inventions of the author. The facts, by far the less important part of the book, were borrowed from various sources (particularly from reports by the eminent Gregg, author of *The Commerce of the Prairies*, but without crediting the authors. The reader can look for everything else in the book except the true state of affairs as to the natural conditions of Texas.

The groundwork for an accurate and comprehensive knowledge about Texas is found in the work of the Englishman, William Kennedy, for a number of years British consul in Galveston. In addition to a detailed history, there is also an extensive description of all the natural conditions of the country. Careful critical use of all available sources gives this work particular worth. It must be remembered, however, that the author learned to know only a small portion of the country through personal observations and that he had to rely upon reports of other persons, which caused errors and discrepancies to creep in. Speaking generally, the author omitted facts pertaining to the unfavorable conditions of the country, thus making it appear in too favorable a light. This should not be considered as a criticism of the book. The fact that he himself owned extensive lands in Texas may have influenced him to do this unconsciously.

Of German writings about Texas, two deserve further mention, which although not very extensive, nevertheless are of some worth since they are based upon personal observations during a long residence in the country. The first is a book by G. A. Scherpf (*Texas, Its History, Geography, Natural History and Topography*) which, in the small space of one hundred fifty pages, gives a condensed but substantial account of the history as well as the political and natural conditions of the country. In general, I have verified by personal observation the accounts of Scherpf, although here and there the natural as well as the other conditions of the country are described in too rosy a light.

The second book deserving of mention was written by Prince Carl von Solms-Braunfels, who had traveled through the country over a longer period and who, therefore, was entitled to an opinion. All that was said in a popular style about the climate, soil and natural conditions is proof of the accurate perception of the author. In the same manner, all the remarks with reference to German immigration to Texas are well-founded and of practical value.

Throughout the book it can be seen that the author had the intention to serve his immigrating compatriots to the best of his ability. A weak point of the book is the biased and unjust criticism directed against the Anglo-Americans. Disregarding almost entirely the excellent qualities of the American national character, he often magnified their faults and bad traits.

Faults of individuals are portrayed as national vices at the expense of the truth and the immigrating German is warned not to acquire any American traits of character, although the honest and unbiased friend of the fatherland sadly misses those very qualities and intellectual endowments among the great mass of his German compatriots, which have made the American nation so rich and powerful within a few decades. Only personal resentment on account of unpleasant experiences, which he perhaps provoked to some extent, alone will explain the warped judgment in this particular point of the otherwise sensible author.

Among other writings devoted entirely to this country, many valuable accounts are also found in the various works pertaining to America in general. Many noteworthy essays of which I will have occasion to quote later are also distributed in the various periodicals.

CHAPTER I
LEAVING NEW ORLEANS—MOUTH OF THE MISSISSIPPI—
ARRIVAL IN GALVESTON—THE HOUSES OF GALVESTON

On November 20, 1845, I boarded a steamboat at New Orleans which was to take me to the seaport of Galveston in the Republic of Texas. At this time Texas was still a Republic, although annexation to the United States was near at hand and was actually consummated during the following month.

The sturdy steamboat, *Galveston*, built for seagoing voyages, formed a very favorable contrast to the numerous river boats anchored along the waterfront of the city, for in comparison they appeared as fragile, bulky square boxes of thin boards flimsily thrown together.

The departure was accompanied by the usual tumult and noise which seem especially great in the slave countries. In this babel of voices, English, French, Spanish and Low-German could be distinguished, and to make this conglomeration of voices complete, from a coasting vessel near us we heard Italian sailors, of whom many live here as sailors and fishermen, calling upon Santa Madonna as a witness to the truthfulness of their assertions.

Unlike the boats of the western rivers which often delay their departure for hours, even for whole days, our boat sailed at the scheduled time and soon we disengaged ourselves from the long row of other steamboats and, after reaching the middle of the river and with it the swiftest current, we disappeared so quickly from the emporium of the Mississippi Valley that only the high cupola of the beautiful St. Charles Hotel was visible, which, viewed from the distance, reminds one of St. Paul's Church in London. Soon after this the obliging captain called my attention to the battleground where in January of 1815, General Jackson with four thousand volunteers from Kentucky and Tennessee gained a brilliant victory over eight thousand picked English troops, thereby saving New Orleans from capture.

Three large live oaks on which long, gloomy-looking Spanish moss (*Tillandsia usneoides L.*) hangs significantly like mourning crepe, indicated the place where the English general had his headquarters.

On the forepart of our steamboat, about one hundred negroes, men and women and children, were huddled together; near them stood horses and various kinds of implements. All this was the property of Virginia planters who were immigrating to Texas. In a great portion of the once rich State of Virginia, the constant planting of tobacco has so impoverished the soil that it is no longer productive enough to warrant its cultivation with negro labor. The planters are compelled to sell their estates and they emigrate South, especially to Texas, in order to buy cheap land where they might make more profitable use of their invested capital: the working power of their negroes. The abandoned plantations are usually purchased by people of the Eastern or New England States who, by the use of white labor and more careful cultivation, restore the productivity of the soil. In the course of time the new owners are rewarded with rich yields of corn and wheat. Since agriculture by means of slave labor entails more expense than by means of free white labor, it can be employed profitably only where the soil is rich. This, more than any other factor, will cause the abolishment of slavery in the Northern States and confine it to those parts of the South where, as in Louisiana, the climate always precludes successful competition by white labor.

Shortly after our departure from New Orleans a farmer's pretty young wife, scarcely more than eighteen years of age, discovered that her husband was not aboard ship. When it became certain that he had been left behind, thus compelling her to make the journey alone to the strange land, her countenance revealed the intense emotions occasioned by this discovery. But with quiet determination so characteristic of the American people, she suddenly took courage, resigned herself to the inevitable, and appeared composed throughout the remainder of the journey. A German woman under similar circumstances would hardly have been able to control her emotions so splendidly.

The following morning I arose early in order not to miss the view of the mouth of the Mississippi. This was only twenty-five miles distant when I came upon deck. The width of the river was greatly diminished at this point (the Delta of the Mississippi) since it had branched off at several places. The banks appeared as a narrow fringe, behind which the land, instead of rising, lay lower than the level of the river. Nevertheless, the shore was covered with low trees, especially willows and poplars. The banks proper soon disappeared entirely and the land or swamp—it is hard to determine which would be the more suitable word—lies almost on a level with the river. Looking across the canebrakes, one experiences the peculiar sensation of seeing the ocean on both sides. Only two narrow strips of swampy land remain as borders of the river. At several places, former arms of the delta were chocked with driftwood and trees were growing on this newly formed land.

Lazy alligators were sunning their ungainly bodies on the accumulated driftwood, but plunged into the stream when our steamboat approached. Flocks of waterbirds inhabited the canals and lagoons of the canebrakes. A school of merry dolphins, evidently intent upon getting a taste of fresh water, were bobbing up and down near our boat. A little farther on down we passed Balize, the pilot-station on the left bank of the river. One can hardly imagine a more desolate and unsuitable habitation for human beings. However, the pilots living here receive sufficient remuneration to enable them to spend several months during the summer in Northern States. At this point we passed several large boats bound from Liverpool to New Orleans which were towed by tugboats. Nearly all arriving sailboats are towed from the mouth of the Mississippi to New Orleans, thus enabling them to complete the last lap of the long journey with speed and safety. Still farther on down, the land on the left bank begins to disappear. A lighthouse rises on the right side among the canebrakes—God only knows on what kind of foundation it was built. The right bank continues in the form of a few detached patches of land, apparently just recently formed by an accumulation of driftwood.

After passing these, our stately steamer is riding the waves of the Gulf of Mexico. It seems as if here nature had forgotten to draw a distinct line of demarcation between the firm and liquid elements. While looking back at the mouth of the river through which we had just passed, now barely discernible, I suddenly comprehended why La Salle, the bold adventurer, at the close of the seventeenth century looked in vain on two expeditions for the mouth of the river from the seaward side, although it had long been explored from the north.

In traveling from New Orleans to the mouth of the Mississippi, the conviction forces itself upon the traveler that the whole Delta of the Mississippi, or the land lying south of the Atchafalaya, is the product of alluvial deposits of comparatively recent date. A glance at the map of Louisiana, drawn on a large scale, makes this clearer. Although the quantity of soil which the river carries with it in its course is so considerable that a glassful dipped from it appears muddy and forms a thick deposit, it must have taken an immense period of time to build up this extensive land from the depth of the ocean bed. The English geologist Charles Lyell computes this period as 67,000 years.

During the course of the day, we again saw land and remained in view of it throughout the rest of our journey. This land was the western coast of Louisiana. We could clearly see the numerous columns of smoke rising from the sugar refineries. Toward noon of the following day, the captain pointed out to us a low piece of land as the Island of Galveston. The coast of Texas, as well as the parts of Louisiana bordering it, are scarcely above the level of the sea and, therefore, void of any points of prominence. Years ago, three live oaks standing near the center, the only trees on the island, served mariners as a guide to the entrance of Galveston Bay.

Shortly after this we saw a white streak which indicated breakers at the bar lying at the entrance to the harbor. We steered a straight course toward it, and soon found ourselves in the midst of the turbulent waves. A few moments later we had successfully surmounted this obstacle, and sailed around the northern point of the island which was separated from another strip of land by a narrow channel. Thereupon we landed immediately at the City of Galveston.

In company with a young American, I went immediately to a boarding house which had been recommended to me. Like most establishments of this kind, it was operated by a woman. The quietness of the place contrasted sharply with the noise of the hotels where I was wont to stay and I felt quite comfortable, although the conveniences offered appeared inadequate to a fastidious European traveler. Individual rooms for each guest or boarder were, of course, out of the question; on the contrary, three or four guests shared one room. The parlor was found on the first floor. Here the guests assembled on cool mornings and evenings around an open fireplace. The three regular meals—breakfast, dinner and supper—were simple, but the food was excellent and plentiful. Beefsteak, fish, cornbread, and home-baked wheatbread formed the menu. A negro family, which our hostess had hired for this purpose from a planter, waited on us. Board and lodging cost one and one-half Spanish dollars daily.

On the following day I inspected the city itself. Galveston resembles other North American cities of similar size and age. It has the architecture and straight streets crossing at right angles in common with other American cities. The houses themselves, including the churches and public buildings, are all of frame construction, and the city can boast of only one brick house. This impresses the European as being only a temporary affair, as for example, for the duration of a fair. The light frame construction, however, does not in any way preclude the presence of cozy, comfortable homes. Considering the mildness of the climate, a more solid, substantial structure is not necessary. The streets are unpaved, but this does not cause any discomfort since the soil is sandy and does not become muddy after a heavy rain. The extreme low altitude of the city present disadvantages, for often sudden springtides in the bay inundate the entire city. During such a time, all traffic is carried on by boats.

With reference to rapid growth, Galveston offers as remarkable an example as any other city of the western part of the United States, such as Buffalo and Chicago, which, like mushrooms, shot up over night. In the year 1838, only one or two houses stood in the place, where today a city has expanded to include five thousand inhabitants, a number of wholesale mercantile houses, three churches and several hotels.

Chapter II
Topography, Wildlife & People of Galveston—
Visiting the Texas Navy—Annexation News—
—A Visit to Dr. Ashbel Smith

On the following day I extended my walks beyond the limits of the city. This region, although level and uniform, is not without its natural charms. The small island upon which the city is built lies southwest by northeast and is from two to four miles wide and thirty miles long. Like the other islands extending as narrow strips of land along the entire southern coast of Texas, so is Galveston Island the product of two opposing currents caused on the one hand by rivers emptying into the present bay and on the other by the tide drifting in toward the coast. Its origin explains its surface and general topography. Save for a sand dune, eight to ten feet high, which extends along the island on the Gulf side, it is entirely flat. The land on the opposite side, facing the bay, is scarcely more than a few feet above sea level, and here numerous lagoons and swamps, inundated during flood-tide, extend into the interior. The soil is predominantly sandy and on the western side only, opposite the bay, is found a black mud of decomposing vegetable matter.

The general appearance of the island reminds one of a prairie. Tall grass covers the flat surface as far as the eye can see, and only here and there is this regularity broken by swampy places overgrown with reeds and rushes. There are no trees found on the island except the three live oaks mentioned previously. Of recent date, however, many ornamental trees have been planted in the gardens of the city, particularly the China tree (*Sapindus chinensis L.*) a great favorite in the South, and their rapid growth clearly demonstrates that the soil and location of the island offer no natural obstacles which the absence of native trees would presuppose. A thorny shrub (*Acacia Farnesiana Willd.,*) eight to ten feet high, is frequently found on the northern part of the island. No large game is found. Deer, although reported as plentiful at one time, have long since disappeared, largely because the barren island offered no refuge for them when pursued by the hunter. I did not ascertain whether or not the island also served as habitat for other mammals. However, great flocks of large and small waterbirds inhabit the lagoons and swamps. Among these, the long-billed pelican (*Pelicanus Americanus and P. fuscus L.*) attracts the attention of the stranger from the North. The so-called horned frog (*Phrynosoma orbiculare Wiegm*) member of the reptile family is also an object of much interest to the newly arrived European, since no animal in Europe resembles it. It is a lizard, about a span long, whose broad, stiff, thorny projections on the back of the head give it a peculiar and formidable appearance. These harmless creatures were very plentiful in the gardens of Galveston. A boy, whom I had commissioned in the morning to capture some, brought me a dozen in the afternoon. I did not see any, however, during my entire stay in Western Texas.

On the Gulf side of the island, the shore slopes very gently toward the sea and forms a level beach which extends uniformly along the entire length of the island, forming an ideal promenade. The sand, washed by the waves, becomes so firmly packed that horses' hoofs scarcely leave an imprint. To drive along the beach in the evening in a light cabriolet drawn by a spirited horse and fanned by a cooling breeze, usually coming from the south, affords the inhabitants of Galveston much pleasure. This pleasure is enhanced by the beautiful view of the Gulf where the waves, rolling in ceaselessly, tumble over each other in the broad, white foaming surf. A bath in the rolling surf is equally refreshing.

The only rough places on the level beach are caused by the driftwood which the ocean casts upon the shore and then buries partially in the sand. Most of this driftwood is composed of huge cottonwood tree trunks, which were carried to sea by the Mississippi and, after drifting about for some time, found their resting place here. The poorer inhabitants of Galveston use this wood for fuel. The greater supply of wood, however, comes from the mainland.

The beach has a peculiar charm for the naturalist who finds here, in addition to numerous mussel shells, many representatives of the various families of a semitropical sea fauna, particularly crabs which dig deep holes in the sand and disappear in them with lightning rapidity at the approach of danger; jellyfish as large as plates; and beautifully formed sea urchins. The marine life shows a marked similarity with that found on the coast of the South Atlantic States. This similarity appears to be greater than one would expect, considering the distance and difference in latitude between these points. It must be traced to the same condition which causes the similarity of the continental flora of the entire North American continent.

Springs or natural accumulations of fresh, drinkable water are not found on the island. The greater part of the supply is obtained by collecting the rain water. The vessels used to store water are called cisterns, but this term is not analogous with the idea which it conveys in Europe. They are large, wooden barrels, usually open at the top and protected by a roof against the direct action of the sun's rays. Every proprietor of a house owns a cistern in which the water, running from the roof, is collected. It was never quite clear to me how this water could remain pure and drinkable even during the hot summer months. Cooled with ice, in which manner it is drunk, it made a palatable drink. If, however, through a continued drought, the supply was exhausted, water had to be brought over from the mainland. I recall when the Mainzer Verein was obliged to maintain a number of immigrants on Galveston Island for several months, the cost of procuring water from the mainland amounted to quite a sum.

The reader may ask: what prompted people to build a city on an island where no corn or wheat grows, where there is neither drinking water nor wood, and where cattle can find only a scanty pasturage? Its good location for commerce is the only answer to this question. The harbor of Galveston is the best along the entire coast of Texas from the Sabine to the Rio Grande. It holds this distinction in Texas, but

otherwise is not classed as a very good one, for ships with a
deep draught cannot enter it. The depth of the water at the
sandbar, lying at the entrance, is eleven feet, which allows
only merchant ships of medium size to pass. Large boats find
good anchorage outside of the harbor. The Bay of Galveston,
at the entrance of which the city lies, extends from this point
thirty miles inland. Since only small steamboats can sail upon
it, Galveston is the staple mart through which all imports
and exports pass, to be distributed in the regions adjoining
the bay and the interior. Steamboats also sail from Galveston
up the Trinity River, which is navigable at all times for quite
a long distance. They also sail up Buffalo Bayou (which empties into the bay) as far as Houston. All products of the fertile
Brazos Valley and also those of the Colorado and Guadalupe
are brought to the latter place, and this city in return supplies
these regions with manufactured goods.

Thus Galveston controls the commerce of the greater
part of Texas. Only a small section in the northern part of
the State, for which the Red River forms the logical highway
of commerce, and a section of the west which lies closer to
Matagorda Bay, are altogether or partially independent of
Galveston.

Nearly all cotton raised in Texas, which is by far the most
important export, is shipped to New Orleans through
Galveston or direct to European harbors. The imports of
manufactured goods are also important since here, as in all
the Southern States, the number of artisans are limited to
the most important ones, such as tailors, carpenters, etc. All
clothing, shoes, boots, etc., are imported ready-made from
the North. For this reason the harbor of Galveston is much
more important than the number of inhabitants of the city
would indicate.

The city has experienced such a rapid growth because its
location has proven itself to be very healthful in comparison
with other cities lying on the Gulf of Mexico. Yellow fever,
the terrible scourge which, as is known, makes its appearance in other places on the Gulf annually in the summer,

appeared in Galveston only twice, in the years 1839 and
1842. The mild form in which this disease manifested itself,
left many inhabitants in doubt whether or not the sickness
was really yellow fever. But investigations by a scientifically
trained physician, Dr. Ashbel Smith, proved conclusively
that it was. This is not improbable, owing to the proximity
of Tampico and Vera Cruz on the one side and New Orleans
and Mobile on the other.

The climate of the island is exceedingly mild and pleasant.
Extreme changes in temperatures as they are found in other
parts of the State, do not occur here owing to the tempering
influence of the sea. Although orange trees grow out of doors,
they do not seem to thrive as well as in Louisiana. The cold
northers so peculiar to Texas may be the cause of this. I saw
a tall banana plant (*Musa paradisiacal*) which had borne fruit
the previous year, growing in a German colonist's garden.

The majority of the inhabitants of Galveston are Anglo-
Americans, coming from all States of the Union. The social
customs and habits of living, however, conform to those of
the South. The average education of the inhabitants is on a
level with that found in cities of similar size in Louisiana,
Alabama and the Carolinas. The same holds true of the
morals of the community, although the fact that Galveston
is a seaport has caused hoodlums and adventurers to drift
here.

Next to the Anglo-Americans, the Germans form a great
part of the population. Most of the artisans and small merchants are Germans, but a number of the larger mercantile
establishments are also owned by them. A German hotel
which enjoys a good patronage is assured a future existence.
Since the past year, when the German emigration to Texas
had increased so rapidly, many German emigrants have remained here for some period before going into the interior.
In the course of the last few months about 3,600 had arrived
and of this number 700 were still in Galveston. The latter,
all members of the Mainzer Verein, were maintained here at
the expense of the Verein, since boats were lacking to carry

them to Indian Point, the harbor on Matagorda Bay. From this point they were to be transported to their destination in the interior. The majority of these people were peasants from northern Germany and the Dutchy Nassau. They were housed in buildings rented for that purpose by the Verein. There was plenty of commotion in the interior of these houses or sheds. The various families were crowded into the rooms among their carefully guarded baggage and boxes piled up high. The women were engaged with their house duties, particularly with mending the clothes of their loved ones, which showed the wear and tear of a long sea voyage. A continuous cooking and baking was maintained on improvised hearths in the yard. The whole scene reminded one of an oriental caravan. Everyone was in good humor since food was plentiful. These people had no foreboding of the privations and sufferings awaiting them. Neither did they know that within a few months many a soul who cherished the wish to secure a future free from hardships and privations through industry and toil, would die a premature death in the new home.

During the course of my stay, I made several interesting acquaintances. Among them was Mr. Kennedy, the British consul at Galveston. He is an educated, cultured gentleman and the author of a well-known book on Texas which disseminated almost the first accurate information regarding this country, hitherto so little known in Europe. It is also valuable on account of an accompanying map. Mr. Kennedy was recalled shortly after I met him because, with the annexation of Texas to the United States, the political interest of England in this country ceased and the commercial relations were not important enough to warrant the presence of a special representative. In addition to a home, arranged according to the rules of English comfort, Mr. Kennedy also had a well-kept garden in which peas were blooming, a peculiar sight to one accustomed to the northern winters. Mr. Kennedy informed me that his table was regularly supplied, from December to April, with fresh peas from his garden.

I am also indebted to Mr. Ashbel Smith, a scientifically trained physician of long standing in Texas, for valuable information and many favors. Formerly this gentleman held an important position in the government of the Republic. At one time he was even sent to Paris on a diplomatic mission and remained there for some time in order to gain for the young Republic the recognition and support of France. Later he voiced strenuous objections to the annexation of Texas to the United States, thereby losing his popularity. Now he lives in Galveston, retired to private life. When I first visited this gentleman, I found him living in a tiny one-room house, built of boards loosely thrown together. The furniture consisted of a bed, a small table, two broken chairs and a chest full of books and papers. Papers were strewn on the floor in wild disorder. Mr. Ashbel Smith, a middle-aged man with a sharply cut profile, wearing high riding boots, was sitting on a chest of books when I entered; lying on the bed was another gentleman who was introduced to me as Colonel B, former minister of war to the Republic of Texas. Although the surroundings, in comparison to those in which European statesmen were wont to live, startled me, I soon convinced myself that these simple surroundings did not preclude a many-sided, thorough knowledge and a finished urbanity of manner. During my further stay in Texas, I have seen similar contrasts between the degree of culture of men and the surroundings in which they live, and I have often wondered how well-bred, educated men could for years endure with resignation the simplicity and even crudeness of frontier life, where the most common conveniences were lacking, if they could only secure for themselves a station in life. Of course, the Americans on the whole are not agreed that certain pleasures and conveniences are absolutely indispensable as are their English cousins, who would not think life worth living without the accustomed comforts.

I spent Christmas Eve in Galveston. The customary manner of celebrating it by decorating a tree and exchanging presents appeared to be unknown; however, small groups

gathered socially and observed the day in festive spirit. The
negroes celebrated it with a formal ball, the music of which
resounded late into the night. On the following day an acquaintance of mine invited me to go to Tremont House, the
most prominent hotel in Galveston, to indulge in a glass of
whisky punch. This is the national drink here with which
Christmas is celebrated. A large company, intent upon the
same purpose, was already assembled when we arrived.
Among them were several generals, colonels, majors and
a number of captains. A stranger just arrived from Europe
would conclude that Texas possessed a large standing army
upon hearing these high military titles. I myself knew from
my experience in the Western States not to take these titles
too seriously, since they were often only honorary ones,
based upon election into the state militia or sobriquets ar-
bitrarily conferred upon individuals by the public. A friend
informed me that Texas boasted of at least forty persons
who carried the title of General.

On New Year's Day, I took a walk along the beach. The
weather was delightful, the sky as clear as at home on a beautiful May morning, and the air balmy. I stretched out on the
sand for a long time and looked out upon the endless ocean
and watched the surf rolling in ceaselessly. When, upon my
return to the home of my hostess, I saw the roses blooming
in the gardens and beheld a small orange tree, laden with
golden fruit, growing in the front yard, a feeling of joy and
contentment crept involuntarily over me. I had escaped for
the time being the rigors of winter and the oft extolled advantages and pleasures of a northern winter appeared to me
at this moment in a dubious light.

However, a severe cold north wind, swooping down suddenly in the evening, reminded me that this climate is also
subject to disagreeable changes in temperature during the
winter months. These northers are a peculiar phenomenon
in Texas, which will be often spoken of later and, therefore,
deserve mention here. As a general rule they occur between
the months of November to May, and seldom during the

other months. Their sudden appearance is one of their characteristics. Only a few signs, such as a peculiar cloud formation and the flight of birds southward, indicate their approach to the natives. The stranger is not aware of their coming until he hears their doleful whistle and feels their icy blast. Although the temperature takes a sharp drop, it seldom goes below the freezing point, but remains around 43° F. This is felt keenly, however, as the change from the high temperature, usually preceding the northers, is so sudden. If, for example, the temperature in the morning was 80° F., it is often 43° F. in the afternoon after the appearance of the northers. The force of the wind increases until it has reached its culmination and then it gradually subsides. The sky is usually overcast during the storm, which seldom exceeds three days. The severity of the storm as well as its duration decreases with the approach of spring.

During the second winter of my stay in Texas, north winds blew approximately half of the winter days. The influence these northers exert upon the organic creatures is noticeable immediately in man and beast. The Texas farmer, at the first approach of a norther, quits his work in the field and waits at his fireside until it subsides. The driver unyokes his oxen when overtaken by a norther on his journey and finds warmth at the campfire which he kindles on the leeward side of a thicket. Convalescents have particular reason to guard against this influence of the wind, for exposure to it may cause a chill and a recurrence of malaria fever. Horses and cattle leave the pastures in the open prairie at the approach of the northers and seek shelter in the dense forest along the streams. Oxen often run several miles to find a sheltered place. Without these northers, the Texas flora would be much more semi-tropical in character, and orange trees would thrive just as well here as in Louisiana, where they bear fruit regularly.

On the following day, in company with one of my acquaintances, I visited an officer of the Texas navy. We rowed to a warship of the Texas Republic, anchored outside of the harbor. Upon arriving there, we were hailed according to seafaring usage, but upon clambering on deck, we soon saw that this was only the wreck of a once stately ship. The whole vessel, a corvette, was dismantled and the deck was robbed of all its warlike accouterments, except two rusty cannon. Instead of a full crew, only two individuals, apparently semi-invalids, were visible. The officer whom we came to see was in his cabin, which showed signs of deterioration. He received us very cordially, but seemed to suffer from despondency which his sojourn on this ship, inevitably going forward to its destruction, may have exerted upon his mind.

After a short stay, someone suggested a fishing trip, to which sport the American is just as passionately addicted as is his English cousin. Accordingly, we rowed to a place where the fish were wont to bite, but the sea was too rough. In order not to return with empty hands, we rowed to a wreck, partially buried in the sand near the shore, and gathered here in a short time a basket full of large, excellent oysters from the half-decayed planks. There is quite a surplus of oysters in Galveston and several oyster shops serve them at any time of the day either raw or prepared in various ways—fried, roasted, stewed, etc. The species is very large, a dozen making a good meal. In quality and piquant taste they are inferior to the smaller European varieties.

My companion informed me that the wreck where we had gathered the oysters was formerly a warship of the Texas navy. Thus I received a very bad impression of the Texas navy on this day. In fact, it is composed only of a few sad remains which lie in the harbor of Galveston and are fast nearing their final destruction.

During the War of Independence of the Texans against Mexico, the navy was called into existence since it was deemed indispensable in the defense of the country. Under its able commander-in-chief, Captain Moore, it soon gave promise of commanding a position of respect. On several occasions it demonstrated its usefulness and efficiency, as for example on an expedition to Yucatan, where a treaty was to be signed with the people of Yucatan who were also dissatisfied with Mexican rule. But when, later, the danger of an invasion by the Mexicans became remote, no provision for its maintenance was made, and thus it deteriorated to its present condition. Had the vessels been disposed of in time, the State would have realized a handsome profit from their sale.

On January 4, 1846, the steamboat *Alabama* arrived from New Orleans with the news of the almost unanimous decision of the Congress at Washington to annex Texas to the United States. Although this was a foregone conclusion, the news still made a profound impression upon the inhabitants. An important chapter in the history of the American continent thus came to a close. The United States, inhabited by the Anglo-Saxon Race, extended its borders to the Rio Grande, thus coming in direct contact with the Republic whose inhabitants are of Spanish extraction.

The Mexican war which soon followed and which will in all probability be only the first of a series of happenings, the ultimate results of which will be the extension of the boundary of the United States as far as the Isthmus of Panama, has quickly justified the conviction of the importance of this event.

Much has been argued in Texas, and outside of the State, about the advantages and disadvantages of annexation to the United States. This argument was futile, since annexation was necessary and inevitable. No small republic, adjoining a powerful and larger one, having the same religion, the same racial characteristics and form of government, can long remain independent. A union must sooner or later take place. This union came quicker than expected, owing to circumstances which made it imperative.

Several reasons can be advanced why annexation was necessary. Texas needed the protection of the United States, as Mexico still had designs of reconquering the lost province. It was profitable for the United States as annexation removed the possibility of having this young Republic come under the sway of European diplomacy.

CHAPTER III
PLANTATION OF COL. MORGAN AT NEW WASHINGTON—
MR. & MRS. HOUSTOUN—EXCURSION INTO THE PRAIRIE—
ASHBEL SMITH'S PLANTATION

When I thought my curiosity was sufficiently satisfied in regard to the first harbor city of Texas, I made preparations for my journey into the interior. My destination was the German colony on the Guadalupe River. To reach it I had the choice of two routes. I could either take a coasting vessel to the harbor on Lavaca Bay and then ride on horseback the remaining distance of one hundred sixty miles to the colony, or take a steamboat to Houston and then travel on horseback to New Braunfels, a distance of three hundred miles. The former route was used by the Mainzer Verein, which had the contract to transport the German immigrants to the colony, as this offered the shorter overland route and was less costly. I myself chose the latter, owing to poor accommodations on the schooner sailing to Indian Point or Lavaca, and to the uncertainty of the duration of the voyage. The longer land route also promised to be more interesting and instructive.

Accordingly I took leave of Galveston on the afternoon of January 12, 1846, and boarded the steamboat *Spartan*. The trip from Galveston to Houston requires on the average twelve hours. The fare, including accommodations in a cabin, and meals was three dollars. The steamboats used for these trips were not built for seagoing voyages as were those sailing between Galveston and New Orleans, but are similar to Mississippi River steamboats. In fact, I learned later that the boats sailing between Galveston and Houston had formerly seen service on the Mississippi.

In addition to myself, two other European travelers were among the passengers, a Mr. Houstoun and his wife, both from England. Mrs. Houstoun was known to me through her very interesting book, *Texas and the Gulf of Mexico, or Yachting in the New World*, published in 1844. It afforded me great pleasure to make her personal acquaintance, so much

the more because in North America and particularly in the
Southwestern States, one seldom meets with educated persons who travel solely to gather information or for pleasure.
Both had also received an invitation from a gentleman on
board ship to spend a few days on his plantation situated
on the bay, and thus I could look forward to the pleasure of
becoming better acquainted with them. This couple, fond
of traveling, had on this occasion come over to Boston from
England on a steamboat, and to this place via New Orleans.
Mrs. Houstoun informed me, however, that a previous trip
to Texas, described in her book, had afforded her a great deal
more pleasure, since they had come over in their own private yacht, commanded by her husband. Such a trip across
the ocean in a ship equipped with all necessities for comfort
and luxury, carrying a complement of more than twenty
men, six cannons for defense, and a physician, furnishes an
example to what extent and on what a grand scale the rich of
England seek some of their pleasure.

In the meantime we had quickly left Galveston far behind
and only a few masts, disappearing gradually against the
horizon, indicated where the city lay. On the other hand, a
very low strip of land became visible on our left. This was
part of the mainland of Texas which rises several feet above
sea-level on both sides of the bay. At times all view of the
coast disappeared entirely, and one imagined one's self on
the open sea. After sailing several hours, we passed Red Fish
Bar, a narrow, continuous reef, partly submerged, partly
rising above the water level, and extending across the entire
bay. This reef had only one passage through which ships of
shallow draft could pass.

This was the only object of interest on this part of the trip
and we were well satisfied when we were set ashore at New
Washington, the home of Colonel Morgan, who had invited
us to visit him. The reader may conclude, upon hearing the
name of this place, that here perhaps was a city modeled
after the capital of the United States on the Potomac. Mr.
Morgan's house was, however, the only one visible for miles.

The place had received its name from speculators in real estate, who thought they had found in it a suitable location for the metropolis of Texas. Since that time it has been carefully indicated on the maps of Texas.

The house of Mr. Morgan lies on the shore of the bay which rises here to an elevation of twenty to twenty-five feet. It is an unornamented, one-story wooden structure of the architectural type common in this part of the South, surrounded by a lawn on which are scattered several red cedar trees (*Juniperus Virginiana L.*) Back of the house is a dense, virgin forest, the tall trees of which are covered with Spanish moss (*Tillandsia usneoides L.*) resembling a grey curtain. Half hidden among the trees stood a number of log cabins constructed of roughly hewn logs, evidently the homes of slaves belonging to the plantation. On the opposite side extended a low, narrow peninsula covered with tall grass on which several hundred head of cattle, divided into numerous groups, were grazing. The cattle in Texas graze out of doors even in winter, and feeding them in barns is not thought of, although most settlers store a quantity of hay and dried corn leaves as a precaution should fires destroy the grass on the prairies. In answer to my question regarding how many head of cattle he owned, my host replied that he thought he had about one thousand. Later I ascertained that it was not uncommon in Texas for individuals to own such a large herd. The natural conditions and climate favor raising of cattle on such a large scale and the European standard, connected with much work and difficulties, cannot be applied here.

From the house one had a good view of the bay which was about two miles wide. Countless flocks of waterbirds, which I had never seen before, inhabited it. In many places the surface of the water was completely blackened by myriads of wild ducks. Whole rows of white swans, resembling silver bands from the distance, clumsy pelicans, geese and various diving birds without number completed the swarms of the feathered denizens. A confused noise, composed of a thousand-fold cackling and screeching rose up from the water as if coming from a huge poultry yard, and this continued unabated throughout the night.

Our host had planned a bear hunt for the following morning. Mr. Houstoun was an ardent hunter who had hunted tigers and elephants in India, and practically every game animal in the various countries of Europe. Mr. Morgan informed us that bear were still plentiful in the neighboring forests, and an experienced guide assured us we would encounter some. However, a heavy rain during the night made the hunt impossible, since the dogs could not trail them. We had to content ourselves with a short excursion into the neighboring prairie. This was to be made on horseback by the whole party as no one walks in Texas if the distance is more than a mile. The horses which had been grazing in the pasture the past few weeks, had first to be caught by the negro slaves. Although they had not been worked for some time, they were, nevertheless, not very spirited as the dry grass in the prairie does not contain much nourishment. The horses were saddled with the Mexican type of saddles with high pommels and high backs, commonly used by the American colonists. Soon the entire company was on its way. We first passed through a dense forest which looked quite virgin, despite the proximity of the dwellings. Uprooted, decaying trees were strewn about everywhere. If a tree happened to fall across the path, it was not removed but the course of the road was changed, according to an American custom of the Northwest. The predominating species of trees were various kinds of oaks, several species of walnut, elm, the hackberry (*Celtis crassifolia Lam*), etc. A heavy underbrush covered the ground and among the bushes were various kinds of evergreen shrubs. With the exception of the latter, all trees were barren as is the case with us in Germany in winter. The dense, grey festoons of Spanish moss hid the barrenness somewhat, at the same time giving the landscape a strikingly peculiar appearance so noticeable to the newly arrived European.

After passing through the forest, I had my first view of a Texas prairie. An unbroken, level, grassy plain extended for miles before us, on which a few islands of trees and shrubs were scattered in irregular order. The oft-made comparison with an English park on a grand scale appeared very appropriate to me.

This region, moreover, was not very attractive; the long grass was yellow and dry, and because of the heavy spring rains, the water had collected in numerous puddles. The only living things we saw on this broad expanse were Canadian geese (*Anser Canadensis L.*), a chattering flock of which were grazing in the wet grass.

Toward noon we returned to the home of our host, since we had received an invitation from a neighboring planter to dine with him. The latter, a relative of Mr. Morgan, lived about three miles distant. All of us, including the ladies, rode to the plantation on horseback, as ox-carts were the only other conveyances used in this region. The women of Texas and the Southwest learn to ride in early youth; it is, therefore, no more hazardous for them than for the men. We passed a place along the way where our host had planted an orange grove at one time. An unusually severe freeze had killed the trees several years ago. Since that time, Mr. Morgan could not muster enough courage to replant them. In fact, I have not seen large orange trees growing out of doors in Texas. Although they grow into large trees and bear fruit regularly in the neighboring State of Louisiana, they cannot be cultivated successfully here.

The house of the planter who had invited us to dinner was situated pleasantly on the high bank of the bay, surrounded by stately trees. It was built of wood in the same simple, unattractive style as the home of Mr. Morgan. Surrounding the house was a porch resting on wooden pillars about two feet above ground. A porch of this kind, at least on one side of the house, is a necessity in every Texas home. On it the occupants of the house spend the greater part of their time in summer, as it affords protection against the direct rays of the sun and at the same time permits the air to circulate freely.

On a section of the porch, which was enclosed with boards on one side only, we were served a dinner of roast wild turkey. This would have been a rather cool place to serve dinner in Germany on January 14, but here we experienced no discomforts, due to the mild climate.

Although the weather was warm during the day, the mornings and evenings were rather cool, therefore, we gathered around a cheerful fireplace and discussed many things which were interesting as well as instructive to a newcomer like me.

Our host, Mr. Morgan, informed us that the chief source of income on his plantation was cattle raising. He delivered a certain number of beef cattle to the market at Galveston annually. The proximity of this market gave him an advantage over other cattlemen living in the interior. He also carried on agriculture with his fifteen slaves and raised, principally, corn. He produced this staple in such quantities that, after he had supplied the needs of his family, of his slaves, and of his cattle, he still had some for sale.

During the past few years he had also experimented with raising sugarcane and the results were very gratifying; the deep, black alluvial soil near his home seemed well adapted to the successful growing of sugarcane.

Since Mr. Houstoun was considering buying a sugar plantation in Texas, the conversation naturally revolved itself around the probable cost of such a plantation and the approximate returns. The average cost of a negro is about $600; his maintenance $30 a year of which sum $20 goes for food and $10 for clothing. Hired slaves can be had for $10 a month. Good uncultivated land, suitable for raising sugarcane, is offered for sale on the Brazos for $10 an acre. The price of a sugar plantation, including fifty negroes and the necessary buildings, would amount to about $50,000. The large capital involved has, so far, prevented the rapid development of this industry. Nevertheless, the production of sugar has been on the increase during the past few years. The most suitable regions for raising sugarcane are the banks of the Brazos at its lower course near the sea, the land near several small streams such as Caney Creek and the St. Bernard between the Brazos and the Colorado, and a few strips of land on the Trinity River. In the course of years many other regions will prove themselves suitable, particularly so in the West, where sugarcane was raised extensively years ago by the Mexicans near San Antonio. The quality of the sugar raised in Texas is excellent and is recognized on the New Orleans market. After a sugar plantation has once been established, the capital investment, according to general opinion, brings better returns than any other investment in the South.

Mr. Houstoun had grave doubts, however, whether he (a British subject) could own slaves. For this reason he thought of importing free labor from Mexico, which reputedly was cheaper than slave labor.

Our affable host informed us also in the course of he conversation that a much larger house and two store buildings filled with merchandise, once stood where today his dwelling stands. These buildings were plundered and destroyed in 1836 when Santa Anna, the perfidious and brutal commander of the Mexicans, whose life unfortunately was spared, had his headquarters here immediately preceding the Battle of San Jacinto.

After breakfast on the following morning, we made an excursion to the other side of the bay. Mr. Ashbel Smith, who had also accompanied us from Galveston, wished to show us his small plantation. A rowboat took us quickly to the other shore where horses were waiting to take us to the farm of Mr. Smith, about two miles distant.

The whole establishment looked rather dilapidated perhaps because the owner did not make his permanent residence there, but had left the farm in charge of an old negro. The manor house was a common two-roomed log cabin, built of partly-hewn logs. The simple furniture consisted of a bed, a table and a few chairs, the seats of which were made by stretching a calfskin tight over them. On the wall near a huge fireplace, in which logs four to five feet in length could be placed, hung an American rifle with a long, heavy barrel, and a shotgun. In the corner of the room stood a tall cabinet, whose contents contrasted sharply with the surroundings. It contained chiefly books which formed a small but carefully selected library. Not only were the Greek and Roman classics represented, but also the best and choicest selections of English and French literature. Similar contrasts, refinement and culture in rough surroundings containing only the necessities of life, are not unusual in all of the western part of the United States.

Near the manor house stood a small, insignificant house which was inhabited by the negro family in charge of the farm. Several other log houses were visible in the distance, one of which served as a stable and the other as a corn crib. In the front of these buildings was a tract of land containing about thirty acres, enclosed by the usual rail fences. In this cultivated field, corn and cotton were raised, but other crops, such as sugarcane would also have done well here, as the soil was a deep, black humus. There were also rows of peach trees in this field. Of all fruits known in Europe, the peach, which grows in all parts of North America, thrives especially well in Texas. Planting the trees between the rows of corn is a custom universally followed. They yield fruit in such abundance that much of it perishes or is fed to the hogs. When once planted, the trees thrive without cultivation. Often one can recognize an abandoned farm by the peach thickets although the houses and barn have long disappeared.

The level prairie extended in the rear of the farm as far as the eye could see. The view on the other side was more or less limited owing to a continuous forest. Someone suggested an excursion to the surrounding country. This suggestion was immediately carried out, as the horses were still saddled.

In spite of rail fences and isolated farms visible at a distance of from one to three miles, usually half hidden among the trees, the whole region still presented a picture of primitiveness. The cattle which we met were startled upon seeing us and bolted into the thicket like frightened game. Twice a herd of deer (*Cervus Virginianus L.*) burst out of the thicket and sped away in long leaps. A little farther on I saw my first wild turkeys in Texas. We surprised a whole flock of them in a little prairie enclosed by the forest. As soon as they perceived us they fled with necks outstretched, their dark, metallic plumage held close to their bodies. When we were about to overtake them with our horses, they sought safety in flight, disappearing across a deep bayou. (All slow moving streams of the low country near the sea, forming natural drainage canals for the rainwater but having such a slight fall that the tide reaches far inland, are called "bayous" in Louisiana and Texas.)

At this same bayou I saw my first magnolia tree in Texas. It was a large, stately tree about a foot and a half in diameter. Although it was not in bloom at this time of the year, it, nevertheless, presented a picturesque sight. Its dark green, glossy foliage contrasted in a peculiar manner with the festoons of Spanish moss swaying in the wind.

Later I saw magnolia trees on Buffalo Bayou at Houston, but I did not see any in Western Texas. They seem to prefer the moist, warm, coastal climate. I also saw my first fig trees here. Several of them stood near an old farm and one specimen was about twenty feet high and three-quarters of a foot thick. We halted at a farm where Mr. Smith introduced himself as a neighbor. In the course of the conversation the owner asked: "Well, Mr. Smith, you have no white family?" People here also speak of a black family in the same manner, meaning, of course, the slaves belonging to the house. The old Romans also called their house slaves "family." In fact, many conditions and peculiarities of the old Roman slave system find their counterpart in the slavery of the Southern States of North America. For example, the Romans permitted their slaves to possess and manage a limited amount of property, and this condition also with the slaves of America, if not in name, at least in fact.

Upon my return to the other side of the bay, I remained two more days at the hospitable home of Mr. Morgan, while the Houstoun couple returned to Galveston on the following morning, preparatory to sailing to Europe from New Orleans. I utilized my time in acquainting myself a little better with the natural conditions of my surroundings.

Mr. Morgan's house lies about twenty-five feet above the level of the sea. The land toward the interior is of equal height and entirely level, while that toward Galveston (the shores of the bay) is scarcely two to four feet high. The surface soil is composed of humus, one to two feet in depth. The subsoil is a red and black clay. In the latter are found shells of a bivalve (*Gnathodon cuneatus Gray*) scattered about singly. On the shores of the bay, but several feet above the present level, one can see mounds, a fathom deep, composed principally of shells of this mussel. These were evidently deposited at an earlier date when the sea-level was quite different from that of today. This same mussel lives today in the waters of the bay, but not in the narrow inlet which in reality is to be considered the mouth of the San Jacinto and on whose shores the accumulation of shells is found. It is only found where the bay enlarges itself and the water becomes brackish through a mixture of fresh and salt water. These facts prove that the original formation of land in the vicinity of the bay is of comparatively recent origin. Although the relative position of land and sea have changed, the climatic and other natural conditions have not been materially altered.

CHAPTER IV

A Trip up Buffalo Bayou—Capitol Hotel—
Description of Houston—Departure for New Braunfels

On the morning of January 17, I heard the popping noise of the high pressure engine of a steamboat coming from Galveston. It landed at a signal from us and a few moments later after a hasty but hearty farewell from my affable host, I found myself on the way to Houston which is, next to Galveston, the most important city of Texas. The trip offered nothing of interest. The bay became narrower and the water shallow. We entered Buffalo Bayou at the approach of darkness. This prevented me from seeing the beautiful magnolia trees which grow along its bank. When I awoke on the following morning and came upon deck, our steamboat was lying in the quiet water of the river, which was confined to its bed by banks thirty to forty feet high. Several large, dilapidated buildings resembling storage sheds stood on top of the bank. A little beyond them many frame buildings were visible. On the landing place, which was but an incline cut into the bank, a number of negroes with two-wheeled carts, drawn by one horse, were waiting to transport the baggage of the travelers to the hotels.

I turned my trunk over to one of the waiting negroes and started on foot to the "Capitol," which was the highsounding name of the reputedly best hotel in town. As soon as I had climbed the rather steep, slippery incline, I found myself on a straight street. Its width left no room for doubt—I was on the principal street of Houston. The houses were all of frame construction, similar in style to those of Galveston; however, everything looked somewhat dilapidated and less tidy here. Nearly every house on Main Street was a store.

The streets were unpaved and the mud bottomless. An embankment had been thrown up on both sides of the street to serve as a sidewalk, but at the intersections one had to leave this sidewalk and trust one's self to the black mud where bottom was found at a depth of six inches. As a result of this condition, I found justification in the fact that everybody, even the elegantly dressed gentlemen, stuffed their trouser legs into their boots.

I finally reached the hotel which was situated at the extreme end of the street. It was a rather pretentious two-story building, but like most of the houses which I had seen so far, showed unmistakable signs of neglect. The interior was worse and many signs indicated that I had reached the borders of civilization.

On a level with the ground was a rather large room with white walls. In the middle of this room was an iron stove, but otherwise it lacked all comforts, since the furniture consisted of only a few chairs, some of which were broken. A number of men, evidently farmers, clad mostly in coarse, woolen blanket coats of the brightest colors—red, white and green—stood around the stove, engaged in lively conversation. The main topic of their conversation was the news of the annexation of Texas to the United States. In a corner of the room lay baggage of all kinds on the ground. The large brass-lined trunks, in common use throughout America, had as yet not made their appearance here. In their stead, leather saddlebags and, in general, light baggage suitable for traveling on horseback, were in evidence.

A little later my host whispered to me that he had another room in the rear of the house which offered more conveniences. He evidently noticed that I felt rather uncomfortable in these crude surroundings which were not improved by the circumstance that the inmates were continually expectorating tobacco juice. He led me into a rather respectable parlor on the floor of which was a carpet, a rocking chair (the inevitable requisite of American comfort) and—what I considered most important, since it had turned cold in the morning—a fireplace in which a cheerful fire was burning. Also the small number of guests were evidently more refined and outwardly more polished. Instead of the coarse blanket coat, made of woolen horse blankets, the black frockcoat was worn, the universal mark of the American gentleman.

The existence of this room, into which the occupants of the former, as if by agreement, never set foot is after all but an admission, that despite the doctrine of the equality of man promulgated in America, some concessions are made to the man of culture.

As the weather remained inclement, a small company usually gathered around the fireplace during the following days, and from their conversation I gleaned much concerning the country which I was about to visit. Among the guests were several planters from the Brazos who were buying clothing for their negro slaves; furthermore, a judge from La Grange, a small town on the Colorado. This judge was on an electioneering tour as he was running for Congressman to Washington. When he heard that I intended to visit the German colony, he wanted to engage me immediately as his interpreter and asked me to further his cause among my German compatriots. At the same time he expressed a favorable opinion about the Germans, but with what degree of sincerity I was undecided.

I could not definitely ascertain the profession or occupation of several other guests, but I inferred from their conversation that they lived in the western part of the country on the outposts of the settlements, for they talked mainly about the fights of settlers with the Indians in former years and the behavior of the latter at the present time. These vivid descriptions, evidently truthful, had a particular fascination for me, the newcomer to a land whose civilization dates from yesterday. The desire to learn more about those Indian tribes through personal observations, was heightened thereby. The dangers and hardships which the first Texas colonists endured on account of the Indians are altogether similar to those assumed voluntarily by the men who, in the beginning of this century, opened up the huge Mississippi River Valley to the enterprising spirit of the Anglo-Saxon Race. Similar, also, are the indomitable courage and unshakable perseverance with which all obstacles were removed at both places.

Several more travelers arrived from the interior during the
course of the day. They assured us that the whole prairie as
far as the Brazos was under water. The creeks and bayous
also had overflowed their banks and could be crossed only
by swimming. A mounted company had crossed them three
times in this manner and another company five times.

These overflows occur every spring in the coastal region
and disrupt communication with the interior almost entirely, or make travel very difficult. The only means of travel
during such a time is on horseback. Traffic with ox-carts
and horse-drawn wagons ceases almost entirely. To remedy
this situation, elevated roads with ditches on both sides and
bridges at certain points, would have to be built. But who
would be able to build and maintain such roads in a thinly
settled country, where farms are often a half-day's journey
apart? Another chief obstacle, however, is the absence of
road building material.

In the entire level plains of Texas no rock is found. That
found in the hilly parts of the State is not suitable for road
building, as far as I know. However, good, suitable material
is found where the settlements end and the Indian country
begins, i.e. north of San Antonio, New Braunfels and Austin. Up to this time, road building was confined chiefly to
indicating the directions between important points; to fell
trees along the river banks where necessary; to lower the
inclines at the fords; and to install ferries at the larger rivers.

Railroads will form the logical and most natural means
of communication as soon as the country becomes more
settled and prosperous. The wide, level plains and undulating regions seem especially created for them. They will be
the more indispensable, since Texas, when compared with
other States of the Union, has been neglected by nature in
the number of navigable streams.

When I wished to retire at night, I was led into a chamber
where two other beds stood in addition to the one assigned
to me. Both were already occupied by snoring individuals
unknown to me. At first I was inclined to view this as a disad-

vantage of travel in Texas; later I would have been well satisfied if this comradeship had extended itself to one's sleeping quarters only.

On the following day the weather permitted me to become better acquainted with my surroundings. Houston was founded in 1836 immediately after the Battle of San Jacinto which freed Texas from the yoke of the Mexicans. It was named in honor of the celebrated leader of the battle. Next to Galveston, it is the most important city of the State, numbering about 3,000 inhabitants. It owes its rapid growth entirely to its location on Buffalo Bayou which is navigable for steamboats of considerable size. This circumstance will always assure it an important position in the future, no matter what changes occur in the interior.

The city is situated on a level plain partly open, partly wooded, whose level, about sixty feet above the sea, is broken only by Buffalo Bayou. The soil is predominantly black and alluvial, but several places, overgrown with pine trees, are sandy. In addition to the many stores, the numerous saloons in the city drew my attention. Some of them (considering the size of the city) were really magnificent when compared to their surroundings. Upon passing through large folding doors, one stepped immediately from the street into a spacious room in which stood long rows of crystal bottles on a beautifully decorated bar. These were filled with diverse kinds of firewater, among which, however, cognac or brandy were chiefly in demand. Here also stood an experienced barkeeper in white shirtsleeves, alert to serve to the patrons the various plain as well as mixed drinks (of which latter the American concocts many.) Everything was calculated to stimulate the appetite and evidently not without results, for these saloons were always filled. Even the merchant would leave his place of business for a few minutes to indulge in a fiery drink in the nearest saloon. The process of drinking, moreover, does not require much time, as it is carried on standing and for this reason it is repeated five or six times during the course of a day.

Occasionally, one sees a person with a hangdog's physiognomy, which reminds one of the earlier years of Houston, when questionable characters and desperadoes flocked together from all parts of the United States in order to lead a dissolute, unrestrained life, such as is usually found in the extreme outposts of American civilization. The city now enjoys a well-ordered community life and the laws are generally obeyed.

There were also a number of Germans in the city at this time who intended to leave for the interior from here. I met most of them in the home of Mr. Henry Fisher who had lived in Texas a number of years. It was he who had obtained the land grants and colonization rights for the Mainzer Verein in conjunction with Mr. Mueller. The immigrants were housed in one of his buildings. A number of horses, mules and oxen were tied in the yard. The cooking and baking was also done in the yard. The house was packed with people and their baggage. Most of these immigrants were young, educated people, predominantly agriculturalists who were supplied with enough funds to equip themselves. All were engaged in preparation for their journey and filled with joyful hope and anticipation.

Mr. Z., a former agent of the Verein, was also there. He was making preparations to leave for New Braunfels or the colony within a few days, with several wagons loaded with merchandise and a number of people, whose transportation with their baggage he had agreed to undertake for the Verein (of course for remuneration.) Since I deemed it inadvisable to travel by myself, and no opportunity presented itself to travel with a mounted company, I joined this party. The disadvantages of traveling with such a slow-moving train were outweighed by the advantage of having more opportunity to make observations in a leisurely way.

We started on our journey in the afternoon of January 15, 1846. The train consisted of two loaded wagons, one drawn by four, the other by two, horses; three German peasants, a former Prussian ensign whose creditors had made things uncomfortable at home for him and who came to Texas to make his fortune (with what, he did not seem to know himself;) furthermore, a cadet who claimed to have attended a military school in Vienna. This young man, who had lived in this country for several years, was employed as a farm laborer at one time. Later, he learned the cigar manufacturing trade. Now he was full of lofty plans and ambitions, such as most Europeans have when they come to this country, but which they soon lose when faced with the grim realities of a frontier life. Finally, there was Mr. Z, who was at one time superintendent of road-building in Greece, but who was badly rewarded in his native Bavaria on account of his philhellenic endeavors. He therefore had bid old Europe goodbye and had come here to engage in trading and speculation in order to make a fortune. I myself was mounted and, therefore, was able to come and go at will.

We had advance notice that the road to the Brazos was bad, owing to continuous rains. This information was correct. Hardly had we left the city when the flat Houston prairie loomed up as an endless swamp. Large puddles of water followed one another and at several places a large section of land was under water. The long, yellow dry grass and the barren trees added to the drab appearance of the landscape. All of the low coastal region presents a similar picture during this time of the year. Anyone seeing it during the winter only would get a wrong impression of the country. It has happened that persons who had the intention to settle in Texas became so discouraged upon seeing the sad picture of the Houston prairie during this time of the year, that they returned immediately to Galveston and left indignantly, concluding that all of Texas was like this region and that the reports of the natural beauty of the State were misrepresentations.

The wagons sank deeply into the mud, compelling the horses to pull them slowly, step by step. Often the wagons became so mired that it required the help of members of our company to push them out of a bog. We moved forward in this manner. Darkness fell and still we had not reached the end of the prairie, nor did we find a dry place to lie down. Impatient over the delay, I rode ahead for about an hour on the dim trail now barely discernible due to the darkness. Just as I was about to give up hope of finding a better camping place, I saw a fire in the distance and this revived my hopes of finding a camp, and also human beings. Soon I heard the tinkling bells of grazing oxen and upon riding up a small incline, I found a group of men camping under the trees near a big fire. Several prairie schooners, shaped like boats, stood nearby. The men were American farmers from the Colorado who were bringing corn to Houston. After exchanging greetings, they invited me to stay. I unsaddled my horse and sat near them by the well-kept fire. They had eaten their supper but gave a negro boy orders to fry some bacon for me and to bake a few sweet potatoes (*Convolvulus batatas L.*) in the ashes. This order was carried out in short time. Since cornbread was left over and the tin coffee can, standing by the fire, contained plenty of coffee, I had an excellent meal.

After supper I spread out my woolen saddle blanket before the fire and chatted with my new acquaintances. They informed me that this was Piney Point, only nine miles distant from Houston, which was often used as the first stopping place by wagons traveling between Houston and San Felipe on the Brazos. Two hours later one of the wagons arrived; the other was stuck in the mud. It was decided to dislodge it on the following morning after the jaded horses had rested sufficiently.

On the following morning before daybreak, the Americans ordered the negroes to round up the oxen while they themselves prepared breakfast consisting again of bacon, cornbread and coffee. The oxen were hitched—four yoke to each wagon—and urged on by encouraging shouts, the horned beasts of burden were on their way.

All commerce is carried on in this manner in Texas. Many farmers occupy themselves with hauling goods during the winter months when work in the field does not require their presence at home. It is usually a profitable business. During the past few years it developed into a lucrative one when the Mexican War and German immigration laid claim to all available means of transportation. Usually oxen are used, but horses and mules are also employed.

The use of oxen is a decided advantage, as they subsist almost entirely upon the grass found on the prairies. Their broad hooves also do not sink into the mud as easily as do those of a horse or mule. The purchasing price is also much lower, since a yoke of oxen can be bought for from forty to fifty dollars, whereas a pair of good draft horses cost one hundred fifty to two hundred dollars. Of course, the slow pace with which they travel is a decided disadvantage. A heavily laden wagon of 3,000 pounds, drawn by four yoke of oxen, can travel on the average only ten to fifteen miles daily. Driving oxen also requires practice and skill. For this latter reason, the German immigrants who had bought oxen to transport their goods, were often at a great disadvantage. It occurred very often during my stay in Texas, that such immigrants were obliged to abandon their wagon in the middle of the prairies and seek help from the nearest farmer, as the stubborn oxen could not be induced to move, or they had all decamped during the night. The requisite in handling oxen is a knowledge of English commands to which they are accustomed and with which they can be driven more easily than with a whip.

The negroes have a special knack in driving oxen. It has often surprised me to see with what ease and safety twelve-year-old negro boys could drive a heavily laden wagon, drawn by four yoke of oxen, by continually giving commands to the individual animals. It almost seemed to me as if they understood it better than the whites, on account of a certain intellectual relationship that allowed them to see the viewpoint of this horned beast of burden.

CHAPTER V
DESCRIPTION OF THE BRAZOS BOTTOM—SAN FELIPE DE AUSTIN—
THE COLORADO RIVER—COLUMBUS

Shortly after the departure of the Americans, we also proceeded on our journey. We were confronted with the same obstacles met with on the previous day. An extensive, level prairie, now and then broken by a sparse grove of oaks, partially covered with water, lay before us. We moved forward very slowly, occasionally passing isolated and rather miserable looking farms. In the afternoon, we arrived at a larger and more pretentious looking farm, belonging to Mrs. Wheaton, where the road to San Felipe led us over Buffalo Bayou. The latter, a shallow stream in the dry season, was swollen to such an extent that we were unable to ford it with the wagons. This obliged us to make a long detour. Accordingly, we left the road and traveled cross-country. Night overtook us in a wet, open prairie, where not a stick of wood could be found to kindle a fire. It would, furthermore, have been impossible to build a fire since a torrential rain was falling. Dispensing with supper, and after securing the horses, we lay down under the wagons on the wet ground. At daybreak we left this inhospitable place and as soon as we found a dry spot with trees near, we halted and revived our sunken spirits by eating breakfast consisting of coffee, cornbread and fried bacon.

A long, unbroken seam of forest became visible in the distance. This was the wooded bank of the Brazos which we reached in the afternoon. Along the skirts of the forest, we saw several plantations partly hidden among the trees. The first one we reached, belonging to Mr. Gaston, was a rather stately establishment. On the broad gallery of the one-story house stood the prosperous-looking owner. Beyond the house was a row of small log houses, constructed of unhewn logs. These were the cabins of the plantation negroes, of whom there were about thirty. We bought corn and fodder here for our horses. Fodder is the name applied to the leaves

of the corn which are stripped before the corn is fully ripe
and then dried, making an excellent food for horses. We
continued our journey northward, following the fringe of
trees in order to reach again the trail to San Felipe, which we
had been obliged to leave at Buffalo Bayou.

The soil was sandy and dry and the ground began to rise
gently. On the entire following day our road led us through
a pleasant hilly country where the open prairie always remained on our right and the continuous forest on the banks
of the Brazos, on our left. At intervals of three to four miles
we saw isolated farms surrounded by trees which were situated pleasantly on an elevation. In the afternoon, we arrived
at the point where the road from Houston to San Felipe
enters the forest on the bank of the Brazos, but we did not
venture farther, since we knew in advance what difficulties
awaited us.

All rivers in the low parts of Texas have broad, level, usually
densely wooded valleys which were formed by the overflow
of rivers. They consist of a deep rich, alluvial soil, and are
subject to overflows which usually occur in the springtime.
The Americans call such a valley "bottom" which is an indispensable word in giving a brief topographical description of
the regions of Texas, and will, therefore, be used quite often
in the pages of this book.

The bottom of the Brazos in the neighborhood of San
Felipe is about seven miles wide, which compares in width
with the Mississippi at St. Louis. The richest and most fertile soil which Texas possesses, suitable for raising cotton,
sugarcane and corn, lies in the Brazos bottom, particularly
below San Felipe. But it requires great exertion and much
work to clear the land of the growth of trees. In addition to
this, the hated malaria fever thrives especially well here.

The passage of wagons through this bottom during the
wet season of the year is accompanied with great difficulties and delays, and the drivers coming from Houston going westward feel they have passed the most difficult part of
their journey when the Houston prairie and the Brazos bot-

tom are behind them. We were also to experience these difficulties on the following day. Immediately upon entering the forest, the road became almost impassable on account of the bottomless black mud. Tree stumps, which threatened every moment to break the wagons, stood often in the middle of the road. Our wagons got stuck in the mud about twenty times, forcing us to hitch all horses to one wagon in order to get it out of the bog. Often this procedure did not work and then the freight had to be unloaded. This was done at least five times in the course of the day. When complete exhaustion of man and beast compelled us to halt at night, we had gone only six miles and were still a mile from the river. This condition could be remedied easily and quickly by cutting down the trees on both sides of the road, thus widening it. At the present time only the most necessary trees have been felled, opening a narrow, crooked passage for the wagons. The dense, leafy canopy formed by the overhanging trees, excludes the sun's rays and the wind.

The slow progress was not unwelcome to me, as it gave me an opportunity to examine the luxurious vegetation of the bottom in a leisurely way. Throughout Texas there is a vast difference between the vegetation of the wooded bottoms and that of the more elevated prairies. The trees in these localities also differ from each other to a great extent, or if they are of the same species, the manner of growth is dissimilar.

The wooded bottoms are distinguished in general from the forests of the prairies by the more luxuriant growth of trees and by their manifold varieties. The tallest and mightiest trees here, as is the case in the Mississippi Valley, are the sycamores (*Platanus occidentalis L.*) and a variety of poplar (*Populus angulata Ait.*) called cottonwood by the Americans on account of the white wool in which the seeds are imbedded. Several kinds of walnut (*Juglans* and *Carya*), several oaks and elms predominate. A tree, occurring often and deserving mention, is the hackberry which grows to a considerable height.

The climbing vines, which play a more important role here than in the Northern States, are almost exclusively confined to the bottom. Grapevines (*Vitis labrusca L.*), as thick as a man's thigh, stretch from the ground to the highest branches of the trees like huge anchor ropes. A kind of sumac (*Rhus toxicodendron L.*) fastening itself with its tendrils like ivy, also climbs high up the trunks of the trees. In a similar manner the beautiful red trumpet vine (*Bignonia radicans L.*) entwines itself around the trees. Spanish moss also reaches its greatest length and development in the wooded bottoms. It does not appear very often elsewhere except in moist places.

The forests of the prairies are usually composed of several species of oak, elm and several other varieties of trees. The trees do not grow tall and strong here and cannot be compared with those of the German forests. The live oak (*Quercus virens L.*) is usually found isolated or in groups of from five to six. In size and strong growth it does not compare favorably with our German oak, but it is nevertheless a beautiful tree with its dark green, glossy foliage and gnarled trunk. It is seldom seen in the bottoms.

The trees of Texas have the general characteristic of the temperate zone, especially those of the Southern States of the Union. I do not know of other evergreen trees, excluding the coniferous trees, except the magnolia (which is confined to the coastal country) and the live oak.

Only in the bottoms of Texas will the newly arrived European find forests which, in relation to size, are in keeping with an idea formed of a virgin forest; but the term "virgin forest" applied to all forests of Texas is applicable, since the hand of man has wrought very little change in their original state.

On the following morning we were but a short distance from the Brazos. Suddenly we stood at its bank, thirty to forty feet high, composed of yellow clay. Below us the yellow, muddy river flowed in its narrow channel. It appeared to me, in comparison to size and volume, like the Vera or Sieg River. On the opposite bank which rose higher than the left one on which we were standing, the houses of San Felipe were visible. A ferryboat took us and the wagons across, and the only difficulty we experienced was descending the steep bank on the one hand, and ascending the other. This difficulty is found in the middle course of most rivers of Texas, such as the Colorado, the San Marcos, etc. The city of San Felipe de Austin, indicated in large letters on all maps of Texas, is made up of from five to six miserable, dilapidated log and frame buildings. One of these is a combination store and saloon. In front of this building were a half dozen men, who drank whiskey freely, and whose haggard frames and pale faces gave plain evidence of overindulgence in whiskey and of the ravages of fever. Otherwise there were no activities or signs of life visible and we were glad when this dismal, deserted place was left behind us. At one time, San Felipe was supposed to have had six hundred inhabitants but, during the last war between Texas and Mexico when the Texan army retreated, the commander in charge of the Texans ordered it burned to prevent it from forming a base for the Mexicans. Since that time the place has not been rebuilt, owing, probably, to its unhealthful location near the Brazos bottom where huge quantities of vegetable matter are constantly decaying.

Had we followed the customary route, it would have led us to Industry and the German colony on Cummins Creek to La Grange. But the leader of our caravan, becoming impatient at our slow progress, decided to travel a shorter route indicated on the map, in order to make up for lost time. This route was supposed to take us through Columbus on the Colorado, but we had no information whether or not difficulties and obstacles would beset us there.

Accordingly, we drove into the prairie, which begins at San Felipe and extends many miles westward, unbroken by forests and undulating land. We passed several herds of grazing deer (*Cervus virginianus L.*) without killing any as it is impossible to approach them in the open prairie.

We were destined to have poor quarters for the night. Darkness overtook us in the open prairie and at the same time a torrential rain fell, which made it impossible for us to kindle a fire. We sought shelter under the wagons as well as we could.

The weather had cleared up on the following morning and we traveled on through the treeless prairie. Toward noon we came to a stream, the Big Bernard, which was swollen considerably on account of the rain of the previous night. In spite of this, the heavier laden wagon was driven recklessly into the stream. When it had reached the middle where the water came higher than the paunch of the horses, two of them refused to go on and threatened to lie down in the water. The two peasants who were mounted on them, sensing the danger for wagon and people, sprang off and endeavored to separate the horses from the wagon. They succeeded in doing so, after some exertion, and brought the horses back to the shore on which we were still standing. At this moment I witnessed a scene, tragic as well as comical, which I shall never forget as long as I live.

When Mr. Z., the owner of the wagon and the merchandise it contained, saw the precarious position of his wagon, and that it was in danger of being carried away by the stream, his face presented such a picture of despair and fright that it became indescribably comical. He was a rather peculiar person with coarse, marked features, bald head, and spectacles on his nose, dressed in wide linen trousers stuffed into his boots, and wearing a short grey jacket. He was running back and forth on the bank in great excitement, wringing his hands and cursing the luck which had brought him to Texas. Suddenly he took courage, discarded jacket and breeches, and plunged into the stream, keeping on only his spectacles. He immediately began to carry the boxes to the opposite shore. This was accomplished in a short time as the others came quickly to his assistance. The empty wagon was then safely pulled across. The other wagon had no difficulty in crossing the river as it was lightly loaded.

The boxes were opened and all the drenched articles, such as shawls, cotton goods, etc., were spread out on the sandy beach to dry. Since, in the meantime, the sun began to shine and upon closer examination the damage was found to be negligible everybody was soon in good humor again. The warm rays of the sun dried the moist articles quickly, enabling us to resume our journey after an hour's rest. We traveled all afternoon through an open prairie. The grass had been burned nearly all the way from San Felipe to this point. The monotonous black ground extended as far as the eye could see. A few deer which were cropping the tender green blades of grass appearing here and there among the burned stubble were the only living things we saw. At another place several prairie chickens (*Tetrao Cupido L.*) fluttered out of a strip of long yellow grass which had not been burned, owing to the moist soil found there.

Toward evening we reached an oak grove and with it the end of the prairie, which extended about thirty miles from this point to San Felipe. We found a good camping place with drinkable water and after eating our supper we stretched out near the fire without taking time to erect a tent, as the air was balmy and the beautiful starlit sky seemed to make further covering unnecessary. However, heavy drops of rain, falling upon our faces at midnight, awoke us rudely from our gentle slumbers. A terrific storm, accompanied by lightning and thunder, was about to burst over us. We had scarcely time to seek shelter under the wagons with our woolen blankets, when a heavy rain fell.

On the following day we arrived at Columbus. The first house we saw, after leaving San Felipe, was about two miles before we reached Columbus. The unfertility of the soil accounts for the absence of all settlement. Up to the San Bernard it is composed of red sand, loose rubble and kernels of ferric hydroxide evidently belonging to the tertiary formation. From the point where the oak forest begins, extending to the house referred to above, an unproductive, silicious soil, composed of flint rubble mixed occasionally with silicified wood, is found.

Before reaching the house, we were obliged to pass a long, swampy stretch where the wagons had to be partially unloaded and the freight carried across. The house belonged to a small plantation without negro slaves. We bought some hens for one bit (that is twelve and a half cents) apiece and eggs at the same price per dozen. Corn for our horses cost us one dollar a barrel which contained about one hundred ears. According to a custom of the country, to sell or trade anything one possesses, the owner of the farm offered several wolf skins for sale. He had taken them from wolves recently caught in steel traps. The pelts were of various colors—one black, the other yellow, and still another greyish brown. The farmer informed us that such variation in color was quite common among the larger wolves. They were very plentiful in the forest surrounding his house and a number of hogs had been killed by them.

Immediately after passing this farm, the bottom of the Colorado begins, which is only one and a half miles wide at this point. The growth of the trees showed great similarity to those found in the Brazos bottom. Under the trees, among which the sycamores were again conspicuous on account of their height, the ground was covered with impenetrable cane (*Miegia macrosperma Pers.*) fifteen feet high. It did not reach such a height in the bottom of the Brazos.

The Colorado at this point is a rapidly flowing stream, confined to a narrow channel similar to the Brazos. Its waters, as the Spanish name indicates, are yellow and muddy. In relation to its volume and width, it can be compared to the German Weser in its middle course.

We were ferried across and, after ascending the steep incline thirty feet high, we found ourselves in Columbus. The location of this place resembles San Felipe, as the bottom of the river is confined to one side and the prairie begins immediately on the other.

Columbus, by the way, presented a more friendly aspect. The frame houses, numbering eighteen or twenty, had wide porches and stood charmingly in the shade of the old live oaks. Neither were the inevitable requisites of a Texas town lacking, for I counted three stores, two taverns and a smithy. In spite of this there was no evidence that the place was growing, as no new buildings were under construction.

We had gone but several miles on the following morning, in the direction of Gonzales, when one of our wagons ran against a stump, breaking an axle. We were obliged to return to Columbus for repairs, although this was not at all to our liking.

Upon arrival there, we found the place in great excitement. A number of horses were tied to the low branches of the live oak trees near the two taverns; in the entrance hall of the taverns, groups of men were standing, carrying on excited conversations. We soon learned the cause of this excitement: a horse race had just taken place and the result of the race formed the topic of their animated conversation. The national passion and love for racing and, especially, the betting connected therewith, was inherited by the American from his English cousin. The wagers are sometimes quite large, with single bets in a little place like Columbus amounting to as much as five hundred dollars.

We had some difficulty in persuading the blacksmith, who took part in the discussions at one of the taverns and who drank one glass of whiskey after another in the heat of the conversation, to drop his "hippological" discussions in order to repair the damage done to our wagon. When he finally began the work, he continued it with the speed common to the American workman, and in a few hours' time the damage was repaired.

On account of this delay, we were able to travel only four miles out of Columbus. We halted in an oak grove where several Americans, bound for Houston with loads of cotton, were encamped. We spent several hours sitting around an excellent campfire, which the Texans know so well how to kindle and to maintain, and chatted with the teamsters. The tinkling bells of grazing oxen were heard in the distance.

On the following morning we passed through a post oak forest, several miles in width. These forests, which cover a wide area in Central Texas between the Brazos and the Guadalupe, have a remarkable resemblance in winter to the cultivated German oak forests, sixty to eighty years old. The growth of this tree is similar to our German oak, but it is not as sturdy and trees with thick trunks are seldom found. In other forests of North America many varieties of trees are usually found, but in the post oak forests all are excluded with the exception of a few walnuts. Underbrush is also lacking. The soil upon which the post oaks grow is usually of average fertility, but also often sterile and unproductive. In the center of Texas is a wide zone where deposits of gravel and sand are found, and where farming cannot be carried on successfully. Here the land is covered with post oaks. They furnish the most suitable material for the erection of log houses, as the trunks are usually straight and have the desired thickness which is three-quarters of a foot to one foot in diameter. They are also valuable in the construction of fences, as the trunks are easily split into rails.

CHAPTER VI
GONZALES—TYPICAL TEXAS LODGING & MEALS—
SEGUIN—THE FERRY NEAR NEW BRAUNFELS

Toward noon we came to a farm where we made a brief halt to prepare a simple meal. The former Prussian ensign bought the skin of a cougar, or so-called panther (which is commonly corrupted by the American backwoodsman to "painter,") for a half dollar from a negro belonging to the farm, intending to use it as a saddle blanket. The negro told us that these animals were quite plentiful in the neighboring forests. The open prairie began again beyond this house but the ground rose in higher undulations. Here and there a single live oak stood among the tall yellow grass, presenting a view which reminded me very vividly of regions in Southern Germany where fruit trees grow in the midst of ripening grain.

In the evening we reached a little stream called the Navidad, in whose bottom we camped for the night. Only a small seam of woods extended along the stream and on the following morning the trail led us immediately into the open prairie again.

When I use the word "trail," I mean the indistinct, half obliterated wagon track which indicated the way from Columbus to Gonzales. But we had no trouble in following it as it was laid out in a straight course by compass.

On our entire journey we had not seen so many deer as at this place. Mile after mile, we saw them in herds, numbering twenty to fifty head, and at one time we saw at least one hundred fifty grazing like cattle near us. But in spite of this, we, nevertheless, were not able to kill any as they could not be approached unnoticed on the level, open prairie.

Our camp for this night chanced to be in the most desolate, wild region encountered so far on our entire trip. It was on the right bank of a little stream, the Lavaca, and at the same time at the foot of Big Hill, probably the most important chain of hills in Central Texas, rising to an approximate elevation of five hundred feet. The wolves kept up a continuous howl near our camp throughout the night.

At Lavaca, I saw my first cactus. These plants remained our constant companions throughout the rest of our journey. They form a shrub, four feet high, with long oval leaves or jointed stems twelve inches long. On the following morning we climbed the barren summit of Big Hill. A panoramic view lay before us: to the west and south, the low woodland extended as far as the eye could see; the sharp outlines of the first foothills were visible against the horizon in the northwest. I greeted this sight with delight, as I hoped to find there a more worthy field in which to carry on my geological investigations than in the low-lying regions we had just traversed, where the only firm rock found at the river crossings was a loose, coarse calcareous sandstone apparently of recent origin. On the summit of Big Hill were layers of conglomerates, evidently belonging to the post-tertiary period, composed of flint rubble and single pieces of silicified wood held together by a calcareous binder.

Our trail led us several miles along the summit of a range of hills until it descended on the other side and took us into a dense oak forest. A little later it ran into a well-traveled road from La Grange to Gonzales, which is so well-traveled because it forms the connecting link with the western part of the country. We had not seen any houses since leaving the Navidad, but here they again appeared. The poor soil, composed of sand and gravel found especially in the region of Big Hill, may have been the cause of this. But someday good farming land will also be found along this route.

We came to a farm belonging to a German-American who had lived in Texas more than twelve years and seemed to be a decent fellow. His farm, which he had acquired through his marriage to a native American woman, was in an excellent condition. He refused to speak German to us, claiming he had forgotten it. My companions knew that he spoke German, but that he sought to hide this as well as his German origin, presumably since his wife desired it.

It is an unfortunate truth that Germans in Texas, as well as in other parts of the United States, renounce their German origin and think it more honorable to pass as native born Americans. At first I was scarcely able to refrain from expressing my indignation upon hearing of such cases, but upon cooler deliberation I was obliged to confess that there were extenuating circumstances for such an ignominious action.

The poor German peasant or artisan who leaves his native land, in most instances, because it does not offer him the bare necessities of life in spite of hard toil and exertion; and who, in a short time, is assured such affluence for himself and his children in the land of his adoption, which he considered unattainable in the old country; and who, moreover, finds instead of a carefully nurtured discrimination between the upper and lower classes perfect equality with all his fellow citizens; and, who finally, does not miss the German culture and refinement because he did not know them; could not such a person in a feeling of gratitude toward his adopted country, which gives him everything the old one denied, easily come to the point to regard also the people, whose adopted citizen he has become, as superior and amalgamate his characteristics with theirs?

How much more must this be the case when his national consciousness, not strengthened by political activities, is confronted by the energetic and united American nationality, which, due to its virility, has a strong tendency to transform quickly all foreign elements and to assimilate them?

Let us hope that in the future the new-born Germany will imbue its immigrated sons with a higher consciousness of the advantages of the fatherland so that it can send them on their way with a more intense national consciousness. Then (even if a thankful, closer attachment to the institutions of the land which received them so hospitably, will be desirable) the disgrace will cease that Germans will renounce their glorious fatherland and its beautiful language, the standard-bearer of the highest human culture.

The Germans of the educated class in America (with few discreditable exceptions) are not included in this reproach. Conscious of the peculiar intellectual worth of their nation, at the same time also properly valuing the praiseworthy traits of the American character, they have never conceded the superiority of the latter, nor denied their German parentage.

We did not reach Gonzales in the evening, but camped a few miles this side of it in an oak grove. Before reaching this grove, we had to ford Peach Creek which contained scarcely enough water to wet our feet, but which like most streams of Texas, sometimes rises twenty to thirty feet and disrupts communication for several days. On the following morning we reached Gonzales which resembles other so-called cities of West Texas. About thirty to forty poor, dilapidated frame houses and log cabins were scattered about on the level plain. Not far distant, a seam of forest extended along the rim of the Guadalupe bottom. The resources of the place seemed to be in keeping with its cheerless aspect. No sugar, coffee or other necessities could be bought in the entire place—nothing but bad whiskey.

Spring, by the way, had already made its appearance. The peach trees were in full bloom on the day of our arrival, February 3, 1846. In the bottom of the creek we found a suffrutescent variety of chestnut in bloom.

After leaving Gonzales, we followed the course of the Guadalupe until we reached our destination, New Braunfels, the road leading us in agreeable changes through fertile valleys or over low hills composed of gravel and sand. The country was more settled here, as we came upon farms every few miles.

We were ferried across the San Marcos a few miles beyond Gonzales, which is here a narrow, sluggish, muddy stream, scarcely twenty paces wide. Later we learned to know it again in its upper course as a beautiful, rapidly flowing stream of incomparable clearness.

We now came to the farm of Mr. King, an old gentleman with a huge paunch (by the way, in Texas this is a rather rare

attribute). He had come here as one of the first settlers and
in the course of years had developed his place into a thriving
farm. When immigration into Western Texas had increased
rapidly, especially among the Germans, he had found further profitable income by maintaining an inn, which does
not require a great outlay of money. All that is needed are a
few beds for the guests. Almost any farmhouse could serve
as an inn. When the guest arrives in the evening, usually on
horseback, his horse is immediately unsaddled by negroes,
or, in the absence of them, by the traveler himself with the
aid of the host. Thereupon he enters the hall where a bucket
of water, a gourd used as a dipper, and a tin basin are found.
After washing face and hands, the traveler seats himself on
a rather uncomfortable chair, with a seat made of calfskin
stretched tight across it, and chats with the host about poli-
tics or the crops. In the meantime supper is being served in
the living room. In some houses, the host asks his guests in
a sly manner to follow him into an adjoining room, and here
offers them a drink of whiskey or cognac diluted with water
and with sugar added, in order to stimulate the appetite.

Supper consists of tea or coffee, warm cornbread and
fried bacon. These articles of food are always found, but in
the better inns biscuits are served hot in addition to eggs,
butter, honey and canned fruits. The hostess, or at least
some feminine member of the family, sits at one end of the
table and serves the tea. This is done in the most dignified
and solemn manner. The cups are passed in silence, and
later repassed in the same manner to be refilled. No sound
is uttered by her except the necessary question, asked in a
quiet, indifferent tone of voice: "You take tea or coffee, sir?"
"Do you take milk and sugar in your coffee?" In explanation
of the latter question, I wish to remark, that the milk and
sugar are added to the tea or coffee by the hostess serving it.

The host urges his guests now and then to partake of this
or that food, but a conversation on his part does not take
place during the meal. In eight to ten minutes the whole "operation" of eating is finished and the guests assemble on the

porch for an hour, in order to enjoy the cool breezes and to chat before retiring. The sleeping quarters are usually confined to one room where two or three beds are found. Each guest selects his bed and if there is not a sufficient number to go around, the guests must share beds. On the following morning breakfast is served. It is a duplication of supper in every detail, as far as the food is concerned. The journey is then resumed immediately after breakfast.

A lodging of this kind, including corn and fodder for the horses, can be had for $1.00 to $1.25. Considering that everything eaten by man and beast is raised on the farm, with the exception of sugar and coffee, it is apparent that such a business is profitable. All cities and hamlets also have hotels which offer more conveniences at a higher cost.

Mr. King's farm was situated conveniently as well as pleasantly. The house, with its many small out-houses, stood on a hill. Lying in front of it was a small cornfield, forty acres in area, enclosed by a strong fence extending to the bottom of the Guadalupe. Another fence enclosed a thirty-acre pasture, also extending to the forested bank of the river. His farm contained, in addition to this, eighty acres of untilled, unfenced land. His chief source of income was his corn crop; but the raising of hogs, sheep and cattle added to his revenue.

Mr. King did not own slaves, but cultivated his farm with the help of his sons and hired white laborers or slaves. He was trying to sell his farm for $3,000 in order to buy several slaves and establish a new farm elsewhere. The wish to possess slaves is inherent in all Texas farmers who do their own work, since the profitable cultivation of cotton and sugarcane can be carried on only with slave labor. The social standing of a slave-owning planter is also quite different from that of the farmer who has to till his own soil by the sweat of his brow.

It rained very hard the entire night. When, in the morning we had traveled but a mile, a little insignificant creek kept us from proceeding farther, as it was swollen so badly that we could not ford it with the wagons. We were obliged to return to Mr. King's house to wait until the water had receded.

During our extended stay, the young people in the home of Mr. King made us all manner of offers for bartering. One wanted to trade or sell a horse; the other who was soon to be married, wanted to trade a good cow and calf for a black frock coat; a third wanted my saddle with which he had fallen in love, and offered me a much better one in trade, according to his opinion. Boys from eight to ten years participated in the bartering with articles of small value and showed a shrewdness seldom found in boys of the same age in Germany. Trading and bartering are more common in Texas than in any other part of the United States. A Texan is ready at any moment, even while traveling, to trade or sell anything he wears, whether it be his coat or shirt, if he can make an advantageous trade. He expects this from anyone else. He has no conception of becoming attached to an article through constant use and is greatly surprised when a German does not care to part with an article, even if offered a price greater than its worth.

At noon on the following day the water of this brook had fallen sufficiently to allow us to continue our journey. The road led us in pleasant changes through small prairies and forests, with the valley, bordered by a chain of hills, to the right of us and the bottom of the Guadalupe to the left.

The following day brought us safely to our destination. By getting an early start, we arrived at the hamlet Seguin in the afternoon. The houses of this place were half hidden beneath the live oaks, scattered about. Only about a dozen could be seen from the road and they resembled the houses we had seen in other places mentioned. Immediately after passing Seguin, we came into an open, undulating but tree-less prairie which extended along at our right as far as the eye could see. At our left flowed the Guadalupe fringed by trees. On the other side of the river extended another immense prairie.

Darkness had fallen when we finally arrived at the ferry at New Braunfels. The ferryboat was on the other shore and we gave the signal to be ferried across with a horn which, for that purpose, was hanging on an old tree. After our call remained unanswered for some time, someone finally called to us from the opposite shore that we would have to wait until the following morning as the river was too swollen and swift to venture across it at night. This was bad news. Instead of sleeping under a roof, as we had hoped, we were obliged to camp in the open while the rain fell and a cold norther was blowing. This became more uncomfortable, since we knew that shelter was so near. However, we also overcame this hardship and, when the morning dawned, the ferry arrived, guided by two young men with whom I had been well acquainted in the old country. We experienced a peculiar, agreeable feeling to hear German-speaking people again after having heard foreign sounds for so long a time. This feeling was greatly enhanced upon entering the principal street of New Braunfels where we saw German faces, German dress and, in general, signs of German manner of living.

We had consumed seventeen days in traversing a stretch of about two hundred fifty miles in the manner described. One mounted could have made the trip from Houston to New Braunfels in six or seven days easily. The stage which in the meantime has been established, makes it in three and one-half days but, of course, by changing horses at regular intervals. This narrative will suffice to show with what difficulties traveling with heavily laden wagons is connected in a land where no roads are found; and that the mere knowledge of distance does not give the German immigrant the proper guide to gauge the duration and hardships of a trip from the coast to the interior of the country.

I found temporary lodging with two young relatives of mine who had recently come to Texas to engage in agriculture. I, therefore, occupied myself immediately in getting acquainted with my new surroundings.

Chapter VII
New Braunfels—The Guadalupe—The Comal—
Peculiar Dress of the Colonists—
Failure of Educated Germans in Texas

The location and general aspect of New Braunfels is very pleasing, and in all of Western Texas no more beautiful and suitable spot could have been chosen for a settlement. It is at the same time also distinctly different from other places found in Texas or other parts of North America.

The city lies on a small, treeless plain, about one-half mile wide and one and one-half miles long. It is bounded on the south by gently rising hills; on the east by the Guadalupe River; and on the north and north west by Comal Creek. A steep hill, about four hundred feet high, rises just across the Comal and extends in a northeasterly direction. It distinguishes itself on account of its precipitous slope and sharp outlines from the low, gently rounded hills, seen when traveling from Houston across the country. A dense forest covers the slope of this hill, but the trees do not have the verdant green of those found in the bottoms, but are of a dark, olive green color. These are the red cedars (*Juniperus Virginiana L.*) which are found singly among other trees in the lowlands of Texas, but nowhere forming a continuous forest of their own.

The Guadalupe, flowing east of the city, is a stream about thirty paces wide, abounding in water which runs rapidly and tempestuously in its rocky bed, due to its considerable fall. Unlike other streams and creeks of the low coastal country, its water is pure and clear, almost excelling that of an Alpine stream. A narrow strip of forest, scarcely classed as bottom, confines the river to its bed. The Comal, which has an equal volume of water but which excels the former in the clearness of its water and the luxuriant growth of trees on its banks, forms a junction with the Guadalupe above the city. The Comal owes its existence to the confluence of the

Comal Springs and the Comal Creek which takes place near New Braunfels. The unexcelled beautiful Comal Spring has its source at the base of a mountain range, hardly a half mile distant from the city. The Comal Creek has its source about ten miles southwest of New Braunfels in several unimportant springs.

The city, or correctly stated, the hamlet New Braunfels is laid out according to a regular plan. All streets cross at right angles and the principal streets converge at the market square. This plan was not much in evidence when I arrived in New Braunfels, because the houses, instead of adjoining one another, appeared to be scattered at irregular distances over the entire plain. Only the principal street, the so-called Seguin Street, could be distinguished quite well, for although houses were not built on both sides, still the townlots, containing about one-half acre each, were enclosed by fences. Every immigrant present at the founding of the colony had received such a lot.

The houses were of diverse architecture as everyone was allowed to follow his own taste and inclination. Besides, most people had no idea as to which particular type of construction was most suited to the climate. As a result, some houses were of logs, some were of studding framework filled in with brick, some were frame, while others were huts with walls made of cedar posts driven vertically into the ground like the posts of a stockade. The roofs, instead of being covered with the customary wooden shingles found throughout America, were covered with tent canvas or a couple of ox-hides. Most of the houses followed the American style of roofed-in porch. These porches are almost indispensable in a warm climate as they keep the direct rays of the sun away from the interior of the house, as well as afford a cool, airy room for the performance of the various household duties. Most of the houses lacked a fireplace, although a fireplace is a necessity for the American farmer during the cold northers in winter, and no American house is without one. Since most of the houses were built in the summertime, the need for heating was remote. In addition to this, the building of a suitable fireplace required a dexterity which the German colonist did not possess.

At the time of my arrival in New Braunfels, there were probably eighty to one hundred such houses and huts to be found there. How rapidly this number was augmented during the same year I was in Texas, I will have occasion to mention later. Several families were packed into one house, no matter how small it was. The interior of such a house, where men, women and children were cooped up with their unpacked chests and boxes, often looked like the steerage of an immigrant ship.

Upon entering the principal street, a small house attracted my attention upon which three shingles hung, containing the inscriptions "Apothecary," "Dr. K," and "Bakery." Combining an apothecary with the practice of medicine is a very common thing in America, where scientifically trained druggists are seldom found, except a few Germans in the larger cities. These druggists are only drug dispensers without any scientific knowledge of drugs. At first I concluded that the "Baker" was a boarder with the physician-owner of the house, but my companion informed me that Dr. K. actually united in his person the profession of apothecary, baker and physician. What advantages accrued to the public through the combining of these three professions, I am not in position to state, since I did not make use of his pharmaceutical and medical services. However, I learned to know, through daily use, the product of his bakeoven as very good wheatbread. I must conclude, though, that this combination must have been a fortunate as well as lucrative one, for during my stay in New Braunfels, a new, neat and roomy house rose near the old one in which the doctor and his young wife, whom he had chosen from among the daughters of the immigrants, established themselves snugly.

Nothing could prevent one and the same person from discharging the duties of several professions since, in all of North America, freedom of trade exists and no kind of obligation for artisans to join a guild, has yet crossed the Atlantic.

On the same main street is located the Evangelical Church of the place, a moderately large frame building with openings for the windows, but without windows proper. The Mainzer Verein had borne the cost of erecting it. Close by stood a tiny house, the modest home of the Evangelical minister, Pastor Ervendberg who does not perform his spiritual duties with the ease of most of his colleagues in Germany. He receives a very meager salary, which the Verein pays. He must preach on Sunday, teach school on weekdays and furthermore, cultivate his corn field and his garden with the sweat of his brow. I was filled with sincere respect for the man when I saw how untiringly he sought to lead his congregation in diligent work and cheerful endurance of the trials and hardships that are inseparably linked with the first settlement in the wilderness; especially also, how, in the evil days, when virulent climatic diseases decimated the population, in true understanding of his sacred calling, he exercised himself without ceasing to give support and comfort.

Up to this time there was only one house in the market square. On the wooden pillars which supported its gallery, one could usually see horses tied, and on the gallery itself a group of men were usually gathered. This was the chief saloon of the place, belonging to Mr. von C. and Mr. von M., both in the employ of the Verein. Dispensing alcoholic drinks is a very lucrative business throughout Texas, and especially among the German immigrants.

The taste for whiskey, which the German peasants and artisans unfortunately bring with them, is stimulated by the warm climate and further encouraged because the people are deprived of many delightful beverages of their native land, particularly good beer and light wine. They must also do without many former conveniences in the arrangement of the home. Since the purchasing price of these alcoholic beverages is small and the daily wages received are high, it is easily explained why the saloons enjoy such patronage.

Unfortunately, most of the people are not aware of the fact that the indulgence in whiskey in the warm climate of Texas is more injurious than it would be in the cold, moist climate of Germany. The most common and cheapest whiskey in America is distilled from corn. It is similar to our corn-spirits. But the more popular drink with the Americans is brandy or French cognac, of which a small portion only is imported from France, the rest being manufactured in this country. Both drinks were also popular in the saloons of New Braunfels and were drunk, diluted with water, out of beer glasses according to an American custom.

French claret, exported in large quantities to New Orleans, could also be obtained here, and diluted with water made a drink probably most suited to the climate of Texas.

It was customary to have a dance at the aforementioned tavern every Sunday for which a violin furnished the music. Although these dances resembled those of our German village ale houses, the dancers exerted themselves far more, owing to the extreme heat in the small room. Two other frame buildings stood near the tavern—they deserve mention as they were conspicuous because of their size. One of them was the store of Messrs. Ferguson and Hessler. Because of increasing patronage, it was equipped to satisfy the needs and wants of the purchaser, however, at prices three and four times higher than those charged for the same article in Germany. In it were found articles of food and also luxuries; ready-made clothing and shoes; saddles and harness; cotton and silk goods; implements of all kinds and a host of articles classed under the name of fancy hardware.

These stores, moreover, are characteristic of the peculiarities of an American colony, which advances with the whole achievements of civilization and often even with the necessities for refined living in the wilderness. In this manner they surprise and carry it by storm, so to speak, thus bringing about a remarkable contrast between rough primitiveness and the marks of a thousand-year-old civi-

lization, which surprises the European in the forests of
Western America.

The other frame building was a combined hotel, tavern
and store. It was owned by a young German count, H. von
D., formerly an ensign in the Prussian army. He had immigrated to Texas under Prince Solms the year prior to my
arrival here. This young man had developed such practical
shrewdness and business acumen, that in less than a year he
had accumulated several thousand dollars, an accomplish-
ment which not many of his colleagues could duplicate, since
they seldom showed any ability in this respect. The hotel
and tavern, to which he principally owed his success had its
humble beginning at the time of the founding of the city. He
had bought a barrel of whiskey in San Antonio and under a
tent had dispensed it out of the barrel to the immigrants.

Owing to its location, standing on the road between San
Antonio and Austin, the seat of government, and also on
account of the presence of numerous immigrants of the
upper class who had arrived with funds, but who had not
as yet established themselves in their own home, the hotel
did a thriving business. The conveniences offered the guest,
however, were not elaborate, but the affability of the host
was all one could wish for. When the humblest worker, or-
dering a glass of whiskey, was served by the count in person,
one had to come to the conclusion that here, indeed, was
the land of perfect equality and brotherhood. Before closing my enumeration of the public buildings of New Braunfels, I should like to mention a restaurant or eating-house,
the like of which is perhaps not often found in Europe.

At the end of the principal street was a little log house
with one room. The cracks between the unhewn logs were
not filled and the circulation of the outer atmosphere with
the air inside of the building makes the place the more unhindered, as there was an opening for a door but no door.
In the middle of the room was a long table resting on stakes
driven into the ground, for the prairie upon which the house
was built, serves as floor in lieu of an artificial one. Several

rough benches were constructed in the same manner near the table. There was no other furniture in the room, but under the projecting roof of the house, a fire was burning over which pots and pans were hanging. A heavy-set woman of middle age, nicknamed by us the "Dicke Madam," was busy at the fire. At one time she was supposed to have been cook for a petty South-German count, but at the present time owner of this prairie establishment. She had, perhaps, ten or twelve regular guests, young men of the educated class, the majority of whom were former lieutenants, who on account of small wages or a disagreement with their superiors or comrades, or for the love of adventure had turned to the New World. The rest were students who had failed, or former merchants or agriculturists. I joined this party on the first day of my arrival.

This company assembled three times daily at the aforementioned building at eight o'clock for breakfast, at twelve for dinner, and at seven o'clock for supper. In regard to her culinary accomplishments, the "Dicke Madam" had adopted a mixed system of cooking, half German and half American. Coffee, cornbread and beef were the mainstay of our meals; but occasionally we were served wild turkey and venison. In accordance with an American custom, we would receive warm meat for breakfast, mostly beef. The latter was the cheapest commodity, since it could be bought for three cents a pound. The management of the Verein would buy a whole herd of cattle at one time and then slaughter several head daily. The meat was sold to the colonists at cost. All other foodstuffs were very high in price, since they had to be brought from long distances. The cost of a bushel of corn, weighing fifty-two pounds, was one dollar and remained so during my entire stay in New Braunfels.

The clothing of my table companions was a particular object of attention to me. They could hardly have been more fantastic and heterogeneous if they had been taken from the wardrobe of a theatre. The component parts were borrowed from the Indian, the Mexican, the American and the German costumes, but the greater part was a production to suit the capricious taste of the individual.

In addition to the German cloth cap, the covering for the head consisted of a broad-brimmed, peaked Mexican sombrero, or a fantastic fur-lined cap with the long tail of the native grey fox dangling from it. The coat was often made of yellow buckskin, fringed gracefully in Indian fashion, or it consisted of a kind of blouse with sleeves slit almost to the shoulders, fashioned and introduced by Prince Solms. A belt held the trousers in place and in it a pistol, a stiletto or a broad bowie knife was found. Long boots, reaching almost above the knee, were very much in fashion, not because they were suitable to the climate but because they met the requirements of the grotesque and romantic. Shoes and boots, made of soft yellow deerskin, were also worn and, while they were light and comfortable, they were not impervious to moisture. A full beard covered the faces of most men. Spurs, preferably such as are worn by Mexicans, with wheels as large as a dollar, and long projections, completed the outfit.

In cold weather a gay-colored Mexican blanket or, otherwise, an ordinary white woolen blanket was hung about. Sometimes a buffalo hide, rough side turned outward, served this purpose.

I noticed a similar irregularity and romanticism in regard to the clothing of the young German colonists. It seems as if they wanted to compensate themselves in the land of the free for the restraint which the manners and customs of the homeland had imposed upon them. The almost total absence of cultured women also helped to encourage this recklessness in dress.

In addition to my table companions, many educated young men lived in New Braunfels at the time of my arrival. The majority of them belonged to prominent noble families who had come here for the purpose of making their fortune, without first having gained a clear conception in what manner. Most of them had brought with them not only a complete outfit of clothing, weapons and farm implements, but also several hundred dollars in cash. Although the majority of them did not leave home for dishonorable reasons, but for financial ones, hoping to become financially independent here through hard work, I have seen very few reach this goal, but most have ended tragically or in degeneracy, and this during my short stay in Texas.

The cause of this deplorable condition is the fact that farming is the only occupation in Texas which assures an independent living. The colonist who has not farmed in the old country or who had not brought along enough capital to purchase slaves, thus enabling him to carry on agriculture on a large scale as a planter, must have an unusual amount of endurance and willpower. During my year's stay in the German settlements, I have seen quite a few German peasants and laborers, who had come here without any funds, come into possession of little farms through their industry. These supplied them with the necessities of life and even gave promise of future affluence and comfortable living. On the other hand, I have hardly seen ten people of the higher class, supplied with moderate funds, who within a year were able to acquire a house with a fenced-in field and of whom one could hope that they would be able to sustain themselves through their own efforts.

The first requirement for the establishment of a farm is the erection of a log house and the fencing in of a field in which corn can be grown. Most of these young people lose their courage and endurance in the performance of this task. It was humiliating for the majority of them, who had never exerted themselves physically, to use two and three times as much time in felling and splitting trees and with much greater exertion than is required by the ordinary peasant and laborer. They, therefore, decided to have the preliminary work done by hired laborers with the funds still on hand, hoping that when this first difficult task had been accomplished, the remaining work could be done without help.

But difficulties arose here also. The laborers available were few in number, since nearly all found it more profitable to establish themselves temporarily on the parcel of land given them by the Verein than to work for others. Those available demanded very high wages and did very little work in return. Wagons were also lacking to transport the felled trees. Wooden shingles, so necessary to cover the house, were also scarce, as the manufacturing of them required a special dexterity.

The trees suitable for their manufacture were also getting scarce in the immediate neighborhood. Even after having acquired a house and a fenced-in field, a hundred other unforeseen difficulties arose. Draft animals for plowing were lacking or, if on hand, the plowing of the field which had never been touched by the ploughshare, required special training. In addition to this, the colonist had to maintain himself until the harvest with purchased supplies, and in many instances also prepare his own meals as well as do his own washing, since most of them were young, unmarried men.

Very few were qualified to resign themselves to such a life. Soon funds were exhausted and the majority of them remained idle, due to the uncertainty of the future. Idleness and lack of conveniences in their temporary homes drove them into the taverns where the companionship with others, who were in a like predicament, afforded them some kind of consolation. The remaining funds were spent in the taverns; later, the outfit brought along was sold; and, finally, when all help was cut off and further credit denied, some turned to desperate means. A few fought as volunteers in the War against Mexico; others entered the United States Army as privates; still others tried to reach other ports of the United States by way of New Orleans, or returned disappointed to Europe by way of Galveston whenever they could persuade a captain to take them along with a promise to pay the passage money later. May the young men of the educated class inform themselves in the future as to what they may expect. May all who have not the firm intention and the necessary qualifications to establish themselves as farmers, remain away, where instead of the dreamed of good fortune, nothing but disappointment and a tragic end await them.

Chapter VIII
Visiting the Lepans—The Botanist Lindheimer—
Springs of the Comal—Excursion to Mission Hill

On the day of my arrival in New Braunfels, I witnessed a very interesting spectacle. A company of Lepan Indians, passing through on their way to San Antonio, camped near the city on the banks of the Guadalupe. Immediately upon receiving this information, my table companions and I decided to pay them a visit. We crossed the river and soon saw several huts or tents scattered about under the trees. They were constructed of branches and covered with buffalo or deer skin. In the first tent that we visited lay a young Indian of effeminate, Bacchus-like appearance, stretched out on soft buffalo hides, holding a pipe, one and one-half feet long, from which he would take a few puffs from time to time. At his feet squatted a young Indian woman with rather broad, but otherwise not unpleasant features, who gave a little black-eyed two-year-old boy white river pebbles to play with. At the same time she looked up to the young man with the greatest respect, obviously ready to do his slightest bidding. Not far from them sat an old woman with an ugly, wrinkled face, apparently the mother of the young warrior or that of his wife. This whole scene of an Indian family life with the man seemingly taking a domineering position, reminded me of a picture from the orient.

A few paces farther on stood another tent, in the middle of which a little fire was burning. An apparently very old man with long, white hair, dressed scantily in old hides, was cowering near the fire, over which he held his hands trembling from age or with cold. Upon coming closer, we observed that the old man was totally blind. The imaginative mind can hardly picture a more sad, cheerless existence than this old savage led, dragged as he was from place to place by his tribesmen among whom he probably had no near relatives and without seeing the living creatures of which he was a part. And how long may he have traversed the endless prairies in this manner?

More pleasant was the sight of the other tents, some of which were only a wall thrown up as a windbreak. Men and women came to us and offered to sell or trade buffalo and deer hides. They demanded particularly woolen blankets and powder in exchange. A tanned buffalo hide was valued from two to four dollars.

Most of the men were tall and well proportioned. The majority had true Indian features with broad, high cheek bones. Several were conspicuous on account of their large hooked nose, which as a general rule, is not characteristic of the American Indian, however, is also found among the Comanches.

The dress of the men consisted partly of skins of animals, partly of cloth of European manufacture. The following articles of clothing are usually worn by the North American savage:

(1) The so-called breech-clout, i.e., a cloth several feet long and about nine inches wide, which is wound around the loins to protect the abdomen.

(2) The leggins, i.e., a covering consisting of leather reaching from the foot to the middle of the thigh. These are similar to our riding leggins, however, they are not buttoned on the side but put on like a stocking.

(3) The moccasins, i.e., shoes, the majority of which are made of one piece of deerskin and usually decorated in various ways.

(4) A woolen blanket or buffalo robe for use in cold weather to cover the upper part of the body which otherwise is naked. The breech-clout is made of blue or red woolen cloth. The leggins are often made of blue cloth but more frequently out of soft, tanned deerskin. In the latter case, the sides are decorated with long fringes. The moccasins are often ornamented with blue or white beadwork.

The Lepans as well as most Indian tribes of North America wear no hats. The black, straight, glossy hair, in which bird feathers are occasionally fastened for decoration, is the only protection in the hottest sunshine. On account of this, one can distinguish an Indian from a white man even at a distance. The latter would certainly contract fever or other sickness without protection for his head.

The Indian also paints his face more or less with red paint, and wears rings of thick, yellow, copper wire on the wrist and ankles, as well as a band of pearls around his neck.

The dress of the squaws differs somewhat from that of the men. The leggins and moccasins are, however, common to both. The upper part of the body is covered with a kind of short, sleeveless tunic, made of yellow tanned skin, which is decorated with fringes and little yellow pieces of brass. A band of thick lime shells of river clams (*Unio*) made into white bars as long as a finger and perforated lengthwise, is worn around the neck. These are valued highly by the Indians. The women are usually small in statue and thick-set. The face is rather broad, but the features are not at all unpleasant. Black, fiery eyes and rows of pearly white teeth are treasures all possess. Hands and feet are small and daintily formed. The squaws are usually in good humor and laugh at the slightest provocation.

The Lepans in Texas belong to the so-called friendly Indian tribes, i.e., they do not live in open enmity with the white settlers and have permission to roam about within the bounds of the settlements and to hunt. At the same time they are not to be trusted since they do not respect the property belonging to others. They are a small tribe and the number traveling about in the inhabited part of Texas is scarcely more than 300. In language, dress, customs and manner they are closer to Comanches than to any other tribe.

In the afternoon of the same day, the band, numbering about forty, transferred their camp to the other bank of the Guadalupe. The passage itself formed a highly interesting and animated spectacle for us Europeans. The squaws packed all the movable articles, consisting mainly of dried meat in addition to buffalo hides and deerskins, in large bundles of untanned buffalo skins. These they carried to the river, set upon each one a pair of black-eyed papooses and propelled the bundles over the swift, swollen stream to the opposite shore with the greatest dexterity. Two squaws always swam near a bundle.

Upon arriving there, they carried the bundles up the high bank with great exertion but always in the best spirits. Then they cut down little tree branches, out of which they constructed the framework for the tents, which were then covered with skins. The men did not participate in this work, but primped and decorated themselves during this time. After everything had been prepared for their reception, they swam across the stream on their horses.

Later the squaws drove the rest of their many horses across the stream, and in a short time the transfer of the camp was accomplished. In the evening some of the men came to town in full war regalia. Their weapons consisted of bow and arrow, a thin lance eight feet long and a round shield made of tanned buffalo skin. Their horses were small and unattractive, but quick and agile in their movements.

During the first few days of my stay in New Braunfels, I made an acquaintance which proved to be pleasant and valuable to me during my entire visit, and which I now recall with pleasure.

At the end of the town, some distance from the last house, half hidden beneath a group of elm and oak trees, stood a hut or little house close to the banks of the Comal. It furnished an idyllic picture with its enclosed garden and general arrangement and position. When I neared this simple, rustic home, I spied a man in front of the entrance busily engaged in splitting wood. Apparently he was used to this kind of work. As far as the thick black beard covering his face permitted me to judge, the man was in his early forties. He wore a blue jacket, open at the front, yellow trousers and the coarse shoes customarily worn by farmers in the vicinity.

Near him were two beautiful brown-spotted bird dogs, and a dark-colored pony was tied to a nearby tree. The description fitted the man I was looking for. His answer to my question, given in a soft, almost timorous voice, which did not seem in harmony with the rough exterior of the man, confirmed my conjecture. It was the botanist, Mr. Ferdinand Lindheimer, of Frankfort-on-the-Main.

He had acquired for himself an enduring reputation through his many years of assiduous collecting of plants and through his study of the botany of Texas, which up to this time was almost unknown. At one time it had received only passing consideration by the English botanist Drummond, who had traveled hurriedly through the country.

After he had received a scientific education in the best German schools and colleges, where he had also made a special study of the ancient classics, Lindheimer taught school for a number of years in an institution of higher learning. A little more than ten years ago he became dissatisfied with the political conditions of his native land, and this fact as well as the love for adventure drove him across the ocean. In company with companions of like mind he first went to Mexico, and lived there for some time near the charming Jalapa, where he engaged in raising pineapples and bananas. Later he went to Texas and fought as volunteer in the latter part of the War of Independence against Mexico.

At the conclusion of the war he endeavored to make a living through farming and to establish a farm of his own. When this manner of living did not agree with him, he followed the advice of a friend in St. Louis and decided to satisfy the desire of his youth—to make the study of botany his profession. He bought a two-wheeled covered cart and a horse, loaded it with paper necessary to pack his plants, and a supply of the most necessary articles of food, such as flour, coffee and salt. There upon, he sallied forth into the wilderness armed with a gun and no other companions but his two hunting dogs. Here he busied himself with gathering and preserving plants. At times he was solely dependent

upon the hunt for food sometimes not seeing a human being for months. When in 1844, the first great contingent of
German immigrants arrived in Texas under the leadership
of Prince Solms, Lindheimer joined them and was cordially
received by the newcomers, since he was well acquainted
with the country and was also a man of experience. He accompanied them to the Comal.

When, in the following spring, the city was founded, he
asked Prince Solms for an unimportant, worthless piece of
land, situated, however, charmingly on the steep bank of
the incomparable beautiful Comal. He renounced all other
claims and built here the hut described previously and began
to work the rich and largely unknown flora of the surrounding country systematically with a leisure and convenience
hitherto not enjoyed in Texas. He was soon convinced, however, that he could not collect properly and at the same time
meet the demands which housekeeping imposed upon him.
When he came home at night, tired from collecting plants,
he was obliged to prepare his own meals first; when he tore
his clothes in the thick underbrush of the forest along the
river bank he had to take needle and thread and repair the
damage done; when he needed a clean shirt, he had to go to
the river and wash it himself.

He took the proper steps to remedy this situation
thoroughly. He sought and found a companion among the
daughters of the newly arrived immigrants. The hut on the
Comal was roomy enough for two people, and everything
proceeded as anticipated, although with primitive
simplicity. The wife not only takes care of the household
duties, but also thoroughly understands the different
processes of drying plants. She should rightfully share in
the praise which the botanists, who receive Lindheimer's
plants from Texas, heap upon him on account of their
keeping quality and their careful preparation. Accompanied
by Lindheimer, on foot and on horseback, I learned to
know the near and distant surroundings of my new home
through daily excursions.

One of the first of this kind was the springs of the Comal, which lay in a straight course about one mile distant from the city at the foot of the hill previously mentioned. We had to cross the shallow, insignificant Comal Creek to get there. Upon crossing it, we came to a small, but extremely fertile plain on which dense patches of forests alternated charmingly with small enclosed prairies. A road, made by the settlers for hauling the cedar trunks, used in building their homes, from the hilltop, was the only sign of human activity. After traveling some time toward the base of the range of hills, we suddenly heard near us the murmuring of rapidly flowing water, and a few moments later we stood at the most beautiful spring I had ever beheld. The natural basin, about forty feet wide, was of incomparable clearness and on its bottom, aquatic plants of an emerald green color formed a carpet. Low shrubs of the palmetto (*Sabal minor Pers.*) which I had learned to know at a less attractive place, namely in the dismal swamps of New Orleans, lined the banks. An old live oak, decorated with long festoons of grey Spanish moss, spread its gnarled limbs over the basin.

This, however, is not the only spring of the Comal. Near it, in the dense forest, and difficult to approach on account of the steep hill slope, are four or six more springs of even greater volume of water and of equal clearness. Every one of them could turn a mill at its immediate source. All unite nearby and form the Comal which, unlike other streams, does not experience a gradual growth, but is born a sizeable stream.

After sufficiently viewing the springs with pleasurable sensation, we returned, following the course of the Comal. Along this route we had many beautiful views of the river, which surpassed the springs in natural beauty. One place in particular fascinated me, and I have visited it repeatedly, each time with renewed pleasure. This was the only ford along its short course through which a wagon could pass. A narrow, natural gravel dam stretched across it. The water above the dam had a depth of from fifteen to twenty feet, but it was nevertheless of such clearness, that one could see clearly every leaf of the green aquatic plants, covering its bottom. When the sun's rays fell upon it, a magical play of blue and green color was produced. On the lower side of the dam where a dense, virgin forest lines the bank of the stream, the river flows tempestuously and rapidly. The depth of the water is deceptive, owing to its clearness which enables one to see every pebble on the bottom. When viewed from the bank, the water covering the dam appears to be only one foot deep, but in reality it reaches to the paunch of the horses and care must be exercised to avoid being swept away by the rapid stream.

Other points near the city are hardly less beautiful, among them the spot where the Comal Creek forms a junction with the springs and where at the same place a little fall interrupts the course of the stream. Unforgettable also will be the ford, where Comal and Guadalupe form a junction. I have seen nowhere in America nor in Europe more beautiful water nor a more luxuriant growth of trees.

Several other streams of West Texas, such as the San Antonio and the San Marcos, are quite similar to the Comal in that they too issue forth as full-fledged streams from mighty springs. All begin at the foot of the mountain range which crosses Texas in a northwesterly direction and which, as will appear later, is really only a slope of the higher rocky northwestern tableland extending to the lower undulating Texas. This condition is apparently not accidental, but must find its explanation in a common cause, and its solution, of course, offers no great difficulty.

The tableland in general is an arid region whose terrain is composed in many places of fissured, hard limestone beds. The atmospheric precipitation which takes place, sinks through the fissures and crevices of the limestone to the impermeable stone layer, collects in large subterranean channels and breaks out in large springs where the limestone hills end abruptly. The impermeable bed is the clayey cretaceous marl, which outcrops at the surface everywhere

near New Braunfels and apparently extends as the underly-
ing formation under the hard limestone bed of the higher
plateau.

The springs of the Comal have another peculiarity. They
maintain an equal volume of water throughout the year. The
temperature of the water is also always the same, namely
76°F., as I convinced myself through a number of readings
and this temperature perhaps expresses also the average
temperature of Western Texas.

This uniform temperature of the springs, which is not
influenced by the changing weather conditions, is also
maintained by the river (with little variation) in its course,
scarcely two miles in length. The temperature of the air af-
fects the water only negligibly since the river is confined to
its deep bed and contains a great volume of water. For this
reason it smokes like a hot spring during the winter months,
especially during the cold northers when the thermometer
drops to 45 °F. During such a time, the inhabitants of New
Braunfels will always find a warm bath of the most pleasant
kind awaiting them.

On another occasion I made an excursion with Lindheimer
to Mission Hill, which rises on the plain of the plateau lying
north of the city. Our path led us again past the springs of the
Comal, but suddenly ascended the steep, wooded slope of
the hill. The firm layers of a yellowish, white, celicious limestone were visible everywhere. The cedar trees, (*Juniperus
Virginiana L.*) which covered the slopes exclusively, formed
an impenetrable thicket through which a path had to be cut.
The cedars here are not the stunted shrub-like plants found
in the Northern States of the Union, but are stately trees
with straight trunks, seldom more than twenty to twenty-
five feet in height and one and one-half feet thick. They have
a uniformly spreading crown. This cedar forest was a treasure to the colonists of New Braunfels, since the wood was
preferred above all others on account of its durability when
used in building houses and fences. A section of this cedar
forest was destroyed by a forest fire during my stay in New

Braunfels. The fire spread rapidly due to the resinous nature of the wood and the close stand of the trees.

Several varieties of cacti stood in various places where the cedars could not take root owing to the limestone on the surface. They grew in numerous groups; among them were primarily (*O. vulgaris Mill*) two to three feet high, with oval leaves one foot long; then another cactus with different habits (*Opuntia frutescens Engelmann*) with numerous thin branches; and finally a small cactus (*Echinocereus caespitosus Engelmann*) four to five inches high, usually found in groups of three and four.

An evergreen shrub also grew on such rocky places of the slope (*Dermatophyllum speciosum Scheele*; mountain laurel - Lignum vitae of the American colonist) eight to ten feet high, which forms the greatest ornamental shrub of the Texas flora with its glossy leaves and abundance of violet colored, sweet scented flowers which appear in early March. It would be worthwhile to attempt to introduce it in Germany.

As soon as we reached the summit of the hill, the cedar forest ended. An open, grassy plain, only broken here and there by brushwood and scattered live oak trees, spread out before us. It extended to Mission Hill about two miles distant and we had to follow a narrow Indian trail to reach it. Several such trails converge here, leading to Comal Springs. The trail was rocky and broken pieces of black hornstone covered the ground everywhere.

Mission Hill—God only knows to what circumstance it owes its name, since it could not have derived it because of a Spanish mission located here at one time—is a small round hill, covered with shrubs. From its summit one has a panoramic view of the surrounding hilly country, which is almost barren. Only here and there a sparse growth of trees is seen. This view impressed me as being more nearly an original wilderness than any other place I have seen in America. The distant houses of New Braunfels, whose shingled roofs glistened in the sunshine, were the only visible signs of human activities. The fauna was also not well represented on these heights. Lindheimer's dogs chased a few rabbits of the small American species (*Lepus nanus Schreber*) and at another time we saw a black wolf slinking through the high grass.

The difference between the character of this region as compared with the immediate surroundings of New Braunfels and with the undulating one of Central Texas is striking. One can imagine one's self miles distant from there. Instead of the gentle, undulating hills, one can see steep hills and sharp cut hill formations with steep slopes; instead of the fertile black soil, the surface is covered with loose broken stones and at many places the hard, solid rock is visible. The brooks and springs of the lowland are absent entirely; everything is dry. In the deep ravines which are usually as dry as the surface of the soil on both sides, huge torrents of water are precipitated only after a heavy rain.

On this excursion (February 9) I noticed the first signs of awakening spring. On the hill bloomed singly among the rocks a blue, red and white anemone (*Anemone Caroliniana Walt.*); near the springs of the Comal, a blue trandescantia unfolded its blossoms and the shrub of a *Cornus*—a *prunus* species—was blooming already.

On our return trip from Mission Hill, we saw in the distance two mounted men who were coming toward us on the Indian path. But suddenly they wheeled their horses and galloped away with the utmost speed. We could still hear the clatter of the horses' hoofs after they had disappeared under the trees of the steep incline, facing New Braunfels. When they turned their horses, we recognized them to be two young, well-known merchants of New Braunfels. The cause of their sudden flight was made clear to us when several acquaintances met us with the news that several Indians had been seen on the path leading to Mission Hill. From the distance, the rather wild appearance of my companion had given the two young men occasion to mistake us for Indians, and in their fright they did not stop to look at us more closely in order to convince themselves of their error. They were ridiculed greatly on account of their timidity and for having spread such a false alarm, although, as subsequent events proved, the presence of marauding Indian bands so near the city could in itself not have been an impossibility.

At another time we followed the Guadalupe several miles in its downward course. The beautiful stream flows in a deep bed below New Braunfels. A small seam of trees indicates its course. The cypress (*Cupressus disticha L.*) take rank among the latter with their mighty trunks, ten feet in diameter, rising out of the water. In growth and foliage it is very different from the common cypress in Southern Europe. The limbs spread wide, and the light green, dainty foliage is shed in winter. The wood makes excellent building material and is particularly used for shingles in Texas. In Western Texas, this tree grows only along the banks of rivers, and usually so close to the stream that the roots reach into the water.

A natural prairie or meadow one-fourth mile wide, extends between the river and a gently rising chain of hills, on which mesquite trees (*Pleopyrena glandulosa Engelmann*) were scattered. These mesquite trees, which spread also over a great portion of northern Mexico, give to the prairie of Western Texas much of its peculiar character. Since they are mentioned often in the pages of this book, they deserve particular description. The trunk is gnarled and now and then bent, thus making it unfit for lumber. They seldom obtain a thickness of over one to one and one-half feet in Texas, nor a height of more than twenty to thirty feet. Gregg reports that they grow thirty to forty feet high in Chihuahua. They resemble our cultivated acacia (*Robinia pseudoacacia L.*) in manner of growth. The foliage resembles the so-called acacia, inasmuch as it is plumeous. The individual leaves, however, are much narrower and the whole foliage is more graceful and transparent. To find shade under a mesquite tree is like dipping water with a sieve.

After the insignificant blossoms disappear, long, narrow pods appear which resemble those of the acacia. These pods, containing a sweet, sticky substance surrounding the kernel, are supposed to be used by the Mexicans as food for horses. According to Gregg, the Apaches and a few other Indian tribes, grind the pod to a meal from which they prepare the much prized pinole.

The mesquite is of limited value to the colonists, since only the thick, heavy wood is used as fuel, and the saplings for picket fences, in the absence of other wood. However, the presence of mesquite trees is regarded as a sure sign of fertile soil, and one can feel assured without further investigation, that the soil is suited to the growing of corn and other cultivated products. Grass also grows luxuriantly under the mesquite trees and the tender mesquite grass, relished so much by cattle and horses, is usually found where mesquite trees grow. Mesquite trees never form a continuous forest, but are found singly, scattered in the prairie.

They are unknown in the lower coastal country and appear only where the ground begins to undulate. From there they spread over the whole of Northwest Texas. On the way from Houston to New Braunfels, they make their appearance between Gonzales and Seguin. The tree belongs to the mimosa and is the most northern representative of this family in America.

On this same excursion along the Guadalupe, I saw the first specimen of a large chino cactus (*Echinocactus Texensis Hpfr.*) which is distributed in all of Western Texas from the coast to the stony tableland. The cacti draw the attention of the German colonists on account of their singular, stiff form, although not a great variety of them appear in Texas and their size and number are not so conspicuous as is the case in certain parts of Mexico.

People in Europe have the erroneous idea that cacti grow exclusively on rocky, infertile soil. This is not the case in Texas, for very often they are found in the prairie, growing in black, fertile humus, surrounded by high grass.

Although the cactus family in Texas does not contain a great many varieties of species, nor are the individual ones as plentiful as in the tropical lands of the American continent, nor as in Mexico and in the northern half of South

America, still their appearance in Texas does not establish
their northern boundary. On the contrary, a small cactus
grows wild in a sandy place of New York at 42° N.L., and
another kind of the same species was observed by Captain
Bach in the region of the "Lake of the Woods" 49° N.L.

CHAPTER IX
VISITING SAN ANTONIO—THE ALAMO & MISSIONS—
FANDANGOES—COLONEL HAYS & THE RANGERS

On February 19, a more distant excursion was undertaken to San Antonio de Bexar (usually called San Antonio by the American settlers) to which I had looked forward with joyful anticipation. This city, the most important settlement ever founded by the Mexicans on the left bank of the Rio Grande or Rio Bravo del Norte is situated about thirty-one miles southwest of New Braunfels.

We left New Braunfels early one morning. The weather was as balmy and clear as is found in Germany about the first half of the month of May. My companion as well as myself was mounted and each of us carried a rifle slung across the saddle for protection against possible Indian attacks, from which one was never quite safe along this route.

The road led us over an open, undulating prairie of great fertility. A cedar-covered hill slope, similar to that at New Braunfels, was at our right for the first ten miles, which farther on flattened out into a low-lying chain of hills. After a ride of fifteen miles we came to the Cibolo River. The bed of this river was dry, but a little below the crossing was a waterhole where we could water our horses. I have crossed the Cibolo several times since then, but have always found it dry. However, after a thunderstorm, it swells to such an extent that it cannot be crossed on horseback. Farther below, especially where the road from San Antonio to Seguin crosses it, it is a constantly flowing stream.

At the point where we crossed, the bed of the river is cut into white, marly limestone which here forms low, but peculiarly shaped cliffs, in which numerous fossils of the cretaceous formation appeared.

The valley of the Cibolo at this point is a fruitful, level plain overgrown with scattered mesquite trees and at that time it was altogether uncultivated, but its fruitfulness has since induced the establishment of two or three farms. Beyond this

valley, the open prairie extends in gently rolling hills as far
as the eye can see until broken by the forested valley of the
Salado, about six miles from San Antonio.

While watering our horses at this clear brook, we heard,
much to our surprise, military trumpet signals near us and,
at the same time, we saw several mounted men galloping
about. We rode in the direction from which the signals came
and soon saw a military camp in a little level plain, enclosed
charmingly by trees. The white tents were arranged in several rows in perfect order and in the center of each row the
horses were tied to posts placed at regular intervals. This
was the camp of two squadrons of dragoons of the United
States who had already been stationed here for several
months as a protection against the Indians. That troops
can camp out of doors throughout the year, and without
much inconvenience, is a testament to the mildness of the
climate.

The discipline, as everywhere in the American Army, is
very strict; however, the wages in comparison to those of the
European armies, are high. The wage of the enlisted man is
six to eight dollars monthly, and in addition thereto, clothing and equipment which leave no room for complaint.
Officers must be native-born Americans. Among the privates, who are enlisted, however, are many foreigners,
especially Irish and German, since the American is averse
to joining the regular army owing to the restraint imposed
upon his independence. The Germans are preferred to the
Irish by the officers, since, as a rule, they are more adept in
the care of horses and are less addicted to the use of whiskey. We conversed with several such Germans and found
them well satisfied with their lot, however, the monotony
of their manner of living, removed from human habitations
as it was, became irksome to them. A few weeks later, forty
horses belonging to this platoon of dragoons, were stolen by
the Indians from the pasture near the camp, although they
had been guarded by a few men. They were not recovered
despite all investigations.

Immediately beyond the Salado, we entered a thick mesquite brushwood, ten to twelve feet high, which encircles the city of San Antonio for several miles, and which formerly offered the marauding Indians a welcome place of concealment. Along the way stood tall cacti and a stemless Yucca (*Yucca filamentosa L.*) with white, spider-like threads between its long, narrow leaves, which is also to be found in other parts of Texas.

A peculiar species of bird (*Saurothera viatica Lichtenst.*) hitherto not observed by us, appeared several times, running rapidly in front of us along the road. It was about as large as a magpie, but resembled a pheasant on account of its long tail. This, however, is only an apparent resemblance, for it does not belong to the gallinaceous birds, but rather is related to the cuckoo family. The plumage is of a beautiful, metallic green color and each feather is fringed with a white band. The Mexicans call this bird *corre-camino* or *paisano* on account of its habit of running along the road. It is unknown in the eastern part of Texas, probably, because of the absence of the thorny thickets, or chaparrals, which seem to be its favorite haunt. It is, however, found in a large portion of Texas, and Lieut. J. W. Abert of the topographic corps in 1848 observed it at the Canadian River on the road from Santa Fe, New Mexico to St. Louis, Missouri.

Presently we saw the city of San Antonio some distance before us. It lies in a broad, almost level plain. Already from the distance the sight of the city has a somewhat foreign appearance, altogether dissimilar to any other Texas city. The cupola of a large stone building especially drew our attention, since this was an unusual sight in Texas. This foreign appearance became more pronounced when we entered the city itself. After passing a few miserable huts whose walls were constructed by ramming poles perpendicularly into the ground and binding them with strips of raw oxhide, we came to a street with stone houses. This street led us to a rather large square from which several streets diverged at right angles. The square was surrounded on three sides by one-story stone houses with flat roofs; the fourth side was occupied by a church built in Spanish style with a low tower above the entrance and a flat-arched cupola over the chancel.

The entire place gave the impression of decay, and apparently at one time had seen better and more brilliant days. The city, as a matter of fact, which was founded toward the close of the 17th century under Spanish rule and was the seat of the Spanish governors of Texas, was supposed to have numbered ten to twelve thousand inhabitants, while at the present it has hardly seven to eight hundred. About half of these are Americans, and the rest are Mexicans. The former are merchants and laborers. The latter belong to the lower class of Mexicans and their features plainly show a mixture of Indian and Castilian blood. The more prominent Mexican families moved across the Rio Grande at the time of the various sieges and particularly when Texas gained its independence. The Mexicans of San Antonio moreover, are a lazy, indolent race. Neither I nor others who have visited San Antonio for any length of time, have been able to fully understand upon what they subsist, for although one can see small gardens next to the houses, one cannot find any corn or grain fields of any consequence outside the city limits. The fact that they live an extremely temperate and frugal life explains only to some extent how they manage to exist.

Several groups of Mexicans stood around in the square, dressed in their picturesque native garb, with brightly colored woolen blankets wrapped gracefully about them, and wearing broad-brimmed, black hats. Several Americans moved among them, but outwardly they contrasted unfavorably to the Mexicans on account of their slouchy manner and less tasteful frock coat, made of light cotton goods.

We lodged at a hotel which was formerly the home of the Spanish governor. It was built of stone and was rather roomy, with high ceilings. It had but few windows and while these had blinds, they contained no window panes. The walls were plain white and the beams upon which the roof rested were visible. Small cedar boards were laid across these, then followed first a layer of mortar and then a layer of dirt. All roofs of the Mexican houses in San Antonio were constructed in the same manner. Most of the rooms had a small fireplace which appeared tiny in comparison to those found in the houses of the colonists in which can be burned trunks of trees several feet long.

Our host was an American and all the arrangements in the hotel were according to American style. We found no shelter for our horses at the hotel, but were obliged to take them to a livery stable where horses were also kept for hire. We paid one dollar per day for their keep. After supper we took a walk through the city, in order to view the night life which begins at the approach of darkness. I had heard much about the fandangos or dancing amusements of San Antonio as being something unusual. We decided, therefore, if possible to learn more about them. The tones of a fiddle or miserable violin, coming from a house on the square near the church, left no room for doubt where to go in order to satisfy our curiosity.

We entered a long, narrow room, level with the ground, in which two or three wax candles, fastened to the wall, dimly lighted the place. A number of young women, dressed in their best apparel, sat on benches at the end of the room, waiting for an invitation to the dance. They were, however, little noticed for the time being by the men present, as were the violin players who, without interruption, played one and the same monotonous piece of dance music over and over again. At the other end of the room stood a few tables at which games of chance were being played. A motley mixture of Mexicans and Americans crowded around these tables. They were playing Monte, a game of cards to which the Mexican is passionately addicted. The Americans, who are almost equally interested in all games of chance, displayed the same eager interest as did the Mexicans. The piles of Spanish dollars and even gold pieces, lying in heaps upon the tables, formed a noticeable contrast to the miserable appearance of the place and the shabby dress of some of the Mexicans. At one table, an old Mexican woman was banker.

Now and then the play was interrupted, as some of the spectators and a few of the players, usually Americans, turned to the dance. The patiently waiting "beauties," among whom were (to be honest) only a few beautiful faces, although several of them had fiery black eyes and comely figures, were asked for a dance, and the various couples arranged themselves in the proper dance order. The dance which now followed was not a fandango, as one would assume from the name which is applied to these evening gatherings, but a peculiar contradance which moved in rather slow tempo, but despite its simplicity was not an unpleasant sight.

The dancers who, as has already been remarked, were mainly young Americans, indeed sought earnestly to act gracefully toward their partners, but in comparison with the natural charm and grace of their companions, the contrast was very obvious. This national difference was still more apparent when, as happened very seldom, also Mexicans took part in the dancing. In contrast to that dignified movement, peculiar even to the lowest of them, and the charm which is enhanced by the becoming dress, the movements of the Americans seemed clumsy and awkward.

When the dance was ended, every dancer led his partner to a table near the gambling tables. It contained refreshments of the simplest kind, namely a kind of small cake which was sold by an old Mexican woman. Every girl would select some of the pastry and her dancing partner would pay for it. However, as a rule the refreshments were not eaten by the girls, but tied in a cloth which evidently had been brought along for this purpose. Each new dance added another piece to the supply and at the end of the dance, all was carried home, which, after all, was the real purpose for attending the dance.

It should also be mentioned here that these fandangos are visited only by Mexican women of the poorer and uneducated families. The women of the few educated Mexican families who still live in San Antonio, visit these public dance halls as little as do the resident American women.

After watching these peculiar performances for some time, we left the house, for we had learned in the meantime that fandangos were in progress at two other places, and we wanted to see these also. We, indeed, soon found the house pointed out to us. The rooms in which the gambling took place, the gambling tables, the black-eyed señoritas and the long, clumsy frames of their American partners—in short, the whole performance—was so similar to the one we had just seen, that we did not gain much by the change and, therefore, returned to our hotel toward midnight.

Such fandangos take place every night God permits to return and some of the resident Americans have not missed any of them for years. Who furnishes the rooms for these amusements is not clear to me, since the guests do not purchase anything and everyone can enter and leave without paying. However, I presume that the Mexican bankers of the gambling tables find their business lucrative enough to furnish two or three thin wax candles to illuminate the rooms. Another point was not clear to me at first: How can these games of chance flourish here so openly, since the American laws in general, and the laws of Texas in particular, prohibit them under pain of heavy penalties? An American living here for quite a number of years gave me a satisfactory explanation to my query in an unabashed manner which seemed plausible to me after having seen these things with my own eyes. He said, "In San Antonio everybody gambles and naturally there are no accusers and judges, therefore, who are inclined to prosecute." I learned, furthermore, that when the circuit court which periodically holds its session at different places in a certain district, and especially also the grand jury which is compelled under oath to report all knowledge of lawlessness, come to San Antonio, the professional gamblers seek safety by leaving the town and do not return until the court sessions are over.

On the following morning we first visited the ruins of the old fort Alamo, which played such an important role in the War of Independence of Texas. The ruins as well as several other houses are on the left bank of the San Antonio River, whereas by far the greater part of the city is on the right bank.

To reach this place we crossed the river over a wooden bridge which, like the rest of the city, or at least those houses occupied by the Mexicans, showed unmistakable signs of neglect. The river is a beautiful stream as clear as the Comal and of considerable depth and volume.

Since here near its springs, it never rises but maintains an equal stand the year round, the gardens in the rear of the houses extend to the water's edge. At first it almost seemed peculiar to have a considerable stream before me in the direction of New Braunfels without having crossed it previously. The explanation, of course, is found in the fact that we had gone around the springs of the stream, lying two miles north of the city, without noticing them.

It was quite a startling spectacle to see here just above the bridge in the heart of the city, a number of Mexican women and girls bathing entirely naked. Unconcerned about our presence, they continued their exercises while laughing and chattering, showing themselves perfect masters of the art of swimming. Several times a few of them were carried near us by the stream and then they would dive and reappear again quite a distance below the bridge. If this was done to hide themselves from our view, it was the wrong thing to do, for the water was so clear that one could see the smallest pebble at the bottom. My companion informed me that this spectacle was repeated daily and that both sexes of the Mexican population were fond of bathing. No better opportunity for bathing could be found anywhere than the San Antonio River with its crystal clear water of equal volume and uniform temperature both in winter and summer.

The ruins of the Alamo covered quite an area. A circular wall, several feet in thickness, enclosing the whole area, was still in evidence in the ruins. The most conspicuous remains of the ruins were the walls of a stone church, the roof of which had caved in. On its portals the date 1758 could be read, showing it to be of comparatively recent origin, whereas the first general impression one gains of these ruins with their foreign appearance, not in keeping with the style of architecture of the present civilization of the American population of Texas, suggests a more remote period for their beginning. The Alamo was a mission as well as a military post for protection against the Indians. It was supposedly established as a military post in the year 1718 by the Spanish government, while the mission, which had been in existence prior to that time, was in fact the oldest point in San Antonio.

When, in the year 1835 at the beginning of the Texas War of Independence, 250 Texans under the brave leader, Benjamin Milam, took the city by storm after a siege of several weeks, the Mexicans retreated to the Fort Alamo under General Cos. However, they could not maintain themselves here very long and when, in accordance with the terms of capitulation, they were allowed to withdraw their arms, the fort fell into the hands of the Texans.

In the following year, on February 25, the first detachment of the Mexican Army, which Santa Anna had equipped to exterminate the Texans, and which he led himself, appeared before San Antonio. The little Texas garrison, numbering not more than 140 men, withdrew into the fort under its brave leader Colonel Travis. Now began a siege, which, while the number of combatants was small, yet without doubt belongs among the most illustrious examples of heroic bravery and unswerving patriotism recorded in the annals of modern history. Despite the strength of the Mexican besieging army, swelled to the number of 4,000 men, and despite a continuous bombardment directed against the fort, the handful of brave Texans repulsed every assault. Finally, on March 6, being worn out by the constant fighting and vigil, a general assault by the Mexicans caused their glorious death. According to the reliable Scherpf, a white woman and a negro were the only persons of the garrison who left the fort alive.

The victory cost the Mexicans 1,500 men. When the resistance was ended, the garrison consisted of only seventeen Texans, most of whom were sick and bedridden. All others had fallen fighting. In the ruins of the Alamo, Texas has its Thermopolae.

In the afternoon we made an excursion on horseback to the so-called missions which lay several miles below the city on the San Antonio River. These missions, as is known, are establishments which were used by the Spanish priesthood for the purpose of converting the Indians, and at the same time they served in the extension of the political power of Spain, under whose protection they were founded in the northern provinces of Mexico, especially also in California. All were erected to conform to one general plan. As a general rule, high and firm stone walls encircle the square yard, several acres in extent, inside of which lie the necessary buildings, primarily the church, the homes of the priests and of the Indians who were to be converted, and a work shop for carrying on agriculture. The latter, according to a wise plan, was especially used as a means for converting the Indians.

In Texas, the region around San Antonio was the chief point for these missions; there are others, however, at various places in the western part of the State. One of the best known is at La Bahia, or Goliad, at the lower course of the San Antonio River, the ruins of which are still in existence. Another is supposed to have been built between the upper course of the Rio Frio and the Nueces. With reference to the ruins of the mission on the San Saba, I will write in detail later.

The nearest mission—Concepcion—lies about two miles below the city close to the left bank of the river. The road first led us through a ford (the only one in the vicinity of the city where the river can be crossed on horseback without danger) to the other side of the river, past several dilapidated Mexican houses and immediately after to a wooded plain, densely covered with mesquite bushes and trees. This fertile plain, extending for miles, which in the early history of San Antonio was extensively cultivated, is now a complete wilderness.

The artificial canals, through which the plain was irrigated with water from the San Antonio River, are still there.

We had not proceeded very far into this plain when a large stone building met our view. Upon coming closer we saw that it was a church similar in style to the one in San Antonio. A square but not very high tower rose over the main entrance and a flat-arched cupola covered the rear of the building.

Everything was quite well preserved, although large cactus bushes, which grew picturesquely on the cupola, gave evidence that the building had been abandoned long ago. Doors and windows had long disappeared and we could ride into the interior. The walls, blackened with smoke in many places, gave evidence that this place was occupied at various times. In more recent times it served half-wild herds of cattle, which roamed about in the neighborhood, as shelter during the cold northers. Of another building adjoining the church, and no doubt the home of the priests, only the walls and window openings are still remaining.

After we had viewed these ruins sufficiently, and contemplated the strange impression which the stately remains of the stone building, no doubt erected with considerable expenditure of effort, called forth when viewed amid the wholly forsaken and wild surrounding, we continued our way to the second mission, San Jose. This one lies on the right bank of the river about two miles distant from the first. Here the wall encircling the square courtyard is partially intact and, within it, several Mexican families live in miserable huts built out of the ruins of the building. The church itself is of similar architectural style, but built on a larger scale than the Mission Concepcion. The portal of the main entrance is on the west side and is still decorated carefully and tastefully with sculptures which, as traces still indicate, were painted in gaudy colors. In addition to the church, a workshop or magazine is still in existence. The Mission San Jose, furthermore, is no doubt the best preserved of the missions of San Antonio and even a part of the extremely fertile soil formerly belonging to it is planted with corn and fruit by the Mexicans, but of course, in a haphazard way.

The material used in the construction of this building as well as the other missions is composed of two kinds of stone. The one is a light, porous, tufaceous limestone or travertine, which is also found in many parts of Germany, as for example in the area of the Leiner Valley in the vicinity of Goettingen where it is valued highly as a building material on account of its lightness. This stone formation finds its peculiar origin in the deposits of springs containing lime. The cupolas and arched ceilings of the churches in the missions are built out of this material.

The other stone used is a greenish grey limestone, containing clay, which has the peculiar property of being almost soft enough to be cut with a knife when taken from the quarry, but later hardens when exposed to the air. This peculiar minerological product is mentioned in several writings as being found in the region of San Antonio. The limestone, whose geological age can be determined by the numerous fossils, particularly species of the family *Exogyra*, enclosed in it, belongs to the cretaceous formation and is found in several places in the neighborhood of San Antonio. The peculiarity noted above, of hardening after exposure for some time to the air, is simply due to the fact that the water which is enclosed in it mechanically, and which produces a condition of slight mobility among its particles, evaporates and thus makes this limestone especially adapted for sculpturing. For this reason it was used for the sculptured portal of the church of the San Jose Mission.

We also visited the third mission, San Juan, which lies three miles farther down on the left bank of the river. Here the church can be traced only by its foundation walls, but the encircling wall which enclosed the square courtyard, and against whose inner walls several buildings, such as the houses of the Indians, lean, were in almost perfect state of preservation. Several poor Mexican families also dwell here in miserable houses of the inner courtyard, but their presence rather increases than diminishes the impression of desolation and decay which the general view presents.

The fourth mission, La Espada, the smallest of them all, lies several miles downstream on the river. However, we had satisfied our curiosity, and therefore started back home without seeing it, since it does not contain any additional noteworthy features.

No one can view the ruins of the past without being filled with wondering admiration at the energy of those who could erect such pretentious buildings in the wilderness. Whatever one may think of the efforts of the Spanish priests to gain converts, their methods through which they hoped to gain their objective, were elaborate.

The contemplation of these ruins aroused further reflection in me. Is it not rather peculiar that, while the Spaniards put forth such extraordinary efforts to convert the Texas Indians to the Christian religion, the people of North America, on the contrary, who have numerous mission societies and in their pious zeal maintain Christian missionaries in many parts of Asia and Africa at an enormous cost, neglect the Indians of Texas who are so much closer to them. At any rate I did not hear of any American missionaries among the Comanches, or any other tribe, during my stay in Texas. This same noticeable neglect applies also to nearly all the other Indian tribes of North America. Although not wishing to express an opinion as to the desirability of such mission endeavors, it nevertheless appears strange that one should seek to satisfy such pious efforts in a distant land when one's own country offers such a wide field.

As we returned to the city, a long train of more than one hundred pack mules wended its way through the streets. At its head and in the rear rode several wild looking, armed men, dressed almost entirely in leather, with black, broad-brimmed hats, and huge spurs on their feet. Their appearance portrayed that they had traveled a long distance.

This was a caravan of Mexican traders, a so-called "Condueta," just arrived from the Rio Grande. Their object was to carry goods back to Mexico. Such caravans come annually to San Antonio in great numbers, and this trade with Mexico, has given the city its only importance. The chief articles which these traders obtain here are cotton goods and tobacco, which are usually paid for with Spanish silver dollars, because the articles of trade, which they bring along, consisting of Mexican woolen blankets and other things, are usually not in demand. A single caravan sometimes carries with it five to eight thousand dollars in cash. Mexico imposes such high import duties on articles of this kind that the Mexican traders, by smuggling them into the country, make a handsome profit despite the long, tedious journey of several hundred miles through the wilderness.

Quite often these caravans are beset by Indian bands and part of the pack mules and goods are lost. This trade has almost entirely stopped since the outbreak of the Mexican War and it will probably not regain its former magnitude at the conclusion of the war, since a new steamship line has been opened recently on the Rio Grande, which will supply the provinces adjoining the river with the necessary European and North American manufactured goods.

In the evening we again visited a fandango and here met a man, who had gained for himself a wide reputation and a deserved popularity among the settlers of Western Texas as a resolute and successful Indian fighter. This man was Colonel J. Hays of Mississippi, a man who came to Texas as a surveyor, but who, for a number of years, has occupied a successful position in the protection of the western boundary as leader of a company of rangers. The company of so-called rangers which he commands, is composed of about sixty young men, mostly sons of the western colonists, who entered as volunteers and who were formerly paid by the Republic of Texas but since its annexation are now paid by the United States.

Each must furnish his own horse and weapons. The latter consist of a rifle and pistols. Many also carry the Colt's revolving pistol which enjoys quite a popularity in Texas.

The duties of the rangers consist chiefly in making frequent excursions along the borders, so as to keep the Indians in check. Upon news of an Indian raid, or other acts of violence, they are to go in pursuit of the enemy immediately and to mete out punishment. To accomplish their purpose, they live in tents throughout the year, changing the location of the camp from time to time as the necessity demands, but they seldom come to town.

Discipline is almost wholly lacking, but this lack is made up for by the unconditional devotion to the leader, who by example leads all in the privations and hardships they usually endure. No one is punished. The coward or incompetent must face the disgrace of dismissal. A uniform is not prescribed and everyone dresses to suit his taste and needs.

The settlers tell of many audacious fights, of successful surprise attacks and quickly executed marches against the Indians, performed by Colonel Hays and his band. Several times, he, himself, brought about a decisive result through his personal bravery and great dexterity in the use of weapons.

The Indians undoubtedly fear him, and no one is better acquainted through personal observations with the extensive, uninhabited prairies and wild deserts west of San Antonio to the Rio Grande, and the rocky tableland, which begins north of San Antonio and reaches to the springs of the Guadalupe and the Colorado.

I was astonished to find the outward appearance of the man seemingly so little in keeping with his mode of living and the traits ascribed to him. Instead of a wild, material, robust figure, I saw a young, slender built man before me, whose soft, beardless face did not portray his martial occupation and inclination anymore than did the black frock coat in which he was dressed. Only in his flashing eyes could a keen observer see traces of his hidden energy.

Several of his men, who looked wild and reckless, participated in the fandango; among them a former Lieutenant von C. from Hanover, who had joined the company partly due to lack of employment, but also for love of adventure.

He expressed the opinion which to me seemed well founded, that his present military life was quite different from his
erstwhile garrison life in a resident city.

The government of the United States, upon whom the
responsibility for the protection and defense of the country
is now incumbent, since annexation of Texas to the Union,
has organized several such ranger companies since the
outbreak of the Mexican War. They are stationed near the
boundary of the Indian country. It is a proven fact that such
corps, composed of colonists, meet the requirements better
than a detachment of regular troops.

On the following day we started on our return trip to
New Braunfels. The charges at our hotel were one and
one-half dollars daily, and the keep for our horses was one
dollar a day. This seems exorbitant when considering the
three daily meals which were of the simplest kind, and the
fact that we had to share beds and also sleeping quarters
in which were two additional beds, occupied by two unknown and unceremonious guests. However, the victuals
had to be brought a long distance and this accounts for the
high charge.

The thought came to me involuntarily upon viewing the
city and the beautiful fruitful valley from a distance, what an
earthly paradise could be created here through the hands of
an industrious and cultured population. The location of the
city in the broad valley, watered by the beautiful stream and
surrounded by gently sloping hills is most charming. The
climate is delightful and a real winter unknown. We slept
comfortably with open doors and windows on February
15 and 16. One is obliged to heat houses only occasionally
when the northers blow. Figs, pomegranates and all fruits
of the warmer parts of the temperate zone grow here out of
doors. Sugarcane was raised extensively in former years by
the Mexicans and is still cultivated on a smaller scale with
success. The soil is of great fertility and this could be still
more increased by restoring and expanding the irrigation
system used under Spanish rule. The river, due to its consid-

erable fall, offers an important waterpower which could be easily utilized.

The region around San Antonio also has the reputation of being very healthful, and years ago it was not uncommon to see Mexicans a hundred years old. A proverb, widely quoted at one time in Texas, told people who wanted to die not to stay in San Antonio. According to my own personal experience, this proverb seems hyperbolic, for during the summer of 1846, fall-fever, malaria and a virulent form of dysentery, which prevailed in Western Texas, was also found in San Antonio and caused numerous deaths. Probably the concentration of many troops in the city and the surrounding country, destined for Mexico, caused the spread of this disease.

Chapter X
Description of the Flora and
Fauna of New Braunfels

W e completed the return trip in seven hours permitting our horses to walk all the way. On the lonely road we met no one except a Mexican who was bringing merchandise to New Braunfels in his two-wheeled ox-cart. The cart was of the primitive workmanship customarily found among the Mexicans. The wheels had no spokes, but were bulky disks, made of several pieces of wood. Not a piece of iron was used in the construction of the wagon. The driver rode alongside on horseback, dignified and picturesque, wrapped in his bright colored woolen blanket, and from time to time he goaded the oxen with a long, pointed staff.

During my absence a letter had arrived from the Commissioner-General of the "Mainzer Immigration Verein," Herr von Meusebach, whom I had met in Galveston. In this letter the bookkeeper of the Verein, Lieutenant von Coll who had charge of the management of the colony New Braunfels during Herr von Meusebach's absence, was instructed to find quarters for me in the "Verein building" and to accord me all assistance in my scientific pursuits.

By the following day I found myself installed in my new home which offered me far more conveniences than the former and which was the most desirable place I could have found in New Braunfels. All the houses of the Verein officers lay on a hill which arose to a height of eighty feet in the immediate rear of the city. The most prominent house was a one-story wooden building about fifty feet long, whose shingle covered roof supported by pillars, projected on both sides, thus forming a gallery. It contained three rooms; a large middle room or hall and a small room on each side. One of the small rooms was especially assigned to me. The middle room was the assembly hall and dining room and furnished by far the most pleasant resort. It was built in such a manner that the two large folding doors opened to the north and south, thus allowing the almost constant, gentle south winds in summer to circulate freely. The view through the north door which one had of the scattered houses of the city and the forested hills arising in the background was most charming and singular.

Back of the main building was another house which contained the kitchen and the dwellings of several petty officers of the Verein. Near it was another loghouse, the home of the men who had charge of the many horses and mules of the Verein. The horses and mules were kept in a pen made of strong posts.

Across from this pen stood a log house which served as a magazine and warehouse. According to a contract with the immigrants brought over the first year, the Verein had pledged itself to advance provisions from its commissary until the first harvest, as well as to maintain a warehouse where the colonists could purchase all necessities at the cheapest prices. The latter arrangement, which at first proved to be a good one, inasmuch as the Verein as well as the colonists profited thereby, finally ceased to function since the financial situation already at that time did not permit the restocking of the warehouses regularly. The advancing of provisions, however, continued and most of the colonists, numbering about eight hundred at the time of my arrival, were maintained from supplies drawn from these warehouses. Daily, great numbers of men, women and children carrying sacks and other receptacles came to receive their ration or as it was commonly called in military parlance, "to grab." The food distributed thus, consisted regularly of corn, coffee, salt and pork; but in addition to this, small quantities of wheat flour, rice, sugar and dried fruits were also rationed out.

In the immediate rear of the company building a gently undulating entirely open prairie extended to the south as far as the eye could see, which served as a common pasture for the horses and cattle of the residents of New Braunfels.

Among the comforts and amenities of my new home was an extensive library, containing chiefly books on natural science, which von Meusebach had brought with him to Texas and which were at my disposal.

For my collecting trips in the interest of natural history, which were now to begin in real earnest, I bought a mule which in the course of time proved itself a useful and trusty servant, and which accompanied me on all my wanderings in Texas. He patiently allowed himself to be loaded with the manifold objects of natural history. Often he presented a grotesque appearance when I came home in the evening from a collecting trip, carrying in addition to myself, a leather saddlebag full of stones, a bundle of plants, and perhaps a young alligator hanging behind the saddle and a four or five-foot chicken snake suspended from the pommel.

In gathering objects of natural history, I was assisted by practically the entire population of New Braunfels, especially the younger generation, since the peculiar types of animals, unknown in the native land, aroused their interest as much as mine. Nearly every day, birds, snakes, lizards, turtles, fish, etc., were brought to me, and by rewarding the finders with a small coin, I was able to stimulate them to renew their search.

On one of the first days, I received a four-foot specimen of a peculiar fish, which frequents the clear water of Western Texas, the garfish (*Lepidosteus osseus Mitchill*), as it is called by the Anglo-American. This fish is distinguished from other fish by its strong, long snout, resembling a bill, and the rhombic, very hard and firm scales which enclose its body like armor. The gar is predatory and I have often seen them remain motionless in the clear water of the Guadalupe and Comal, evidently awaiting their prey. The first specimen which I opened had a fish a foot long in its stomach. The flesh of this fish is insipid and worthless. Several of the specimens had entangled themselves with the teeth of their long bill in the fishing tackle which had been cast out to catch other fish. Most of them, however, were harpooned from a boat with an iron spear. The scales are so hard that one can open the fish only by separating the borders of the contiguous scales.

On another occasion I received a specimen of a soft-shelled turtle (*Trionyx ferox Say.*) more than a foot long, caught in the Comal. I had difficulty in keeping it as my table companions, who had an Epicurean taste, insisted on sending it to the kitchen. Prepared properly, this turtle furnishes a palatable food and even the soft gristly edge of the shell is eaten. Turtles were occasionally caught near New Braunfels on hooks set out for catfish.

A peculiar twelve-inch crab with very long, thin claws was also caught in the Comal River on the first day I began my collecting. Up to this time it was reported as living in the rivers of Jamaica and as belonging to a species which, with this one exception, lives in the sea (*Palaemon Jamaicensis M. Edwards.*)

On the 18th of March, there came up the hill two yoke of oxen, pulling a wagon which contained a peculiar freight intended for me. It was an alligator almost eleven feet long (*Alligator lueius L.*) which a colonist formerly from Nassau had shot in the Comal Creek about six miles from New Braunfels. In the course of the summer, eight more of varying sizes were shot at a place where the creek enlarged itself into a little pond about thirty paces long. A former student of medicine of the long suppressed University of Herborn, who had lived thirty years in Russia and fought against the Circassins and then immigrated to Texas at the age of seventy, offered to help me in dissecting and preparing the skeleton. We found in the stomach of the monster, 1) bones and hoofs of a deer which seemingly had been there a long time, for they had become flexible and gristly through the action of the gastric juices which had dissolved the lime; 2) a half-digested wild turkey; 3) the bones of a squirrel; 4) several stones and pieces of wood each as large as a fist; 5) a green pulp, seemingly composed of half-digested vegetable matter. Alligators also live in the clear water of the Guadalupe and Comal Springs. They do not cause the colonists much damage, only occasionally devouring a few ducks or a dog which ventures into the water.

Snakes of all kinds were brought to me in great numbers. As most everywhere in Texas, rattlesnakes were plentiful and almost daily some were killed by settlers at work in the field. Not infrequently horses grazing in the prairie were bitten by them. The bite usually caused a painful condition of short duration without causing permanent injury or death. At one time I saw a horse belonging to the Verein, which had been bitten in the mouth a half hour previous. The whole head of the animal was badly swollen and distorted. The eyes bulged out and the animal, no doubt suffering greatly, trembled in its whole body. To bring about relief, several incisions were made in the mouth from which the blood oozed slowly. On the following day I saw this same horse fully restored to health, grazing with the others on the prairie.

Despite all my investigations, I have not heard during my stay in Texas, that the bite of the rattlesnake has caused the death of a person. The best remedy, also used in Texas, for the bite of a rattlesnake as well as other poisonous snakes is spirits of ammonia poured into the enlarged wound. A few drops, diluted with water, should also be taken internally. When traveling in Texas, it is advisable to carry a little bottle of this fluid.

The fat of the rattlesnake, of which the old ones have large quantities, is prized as a fine lubricant, especially for the locks of guns. That hogs destroy rattlesnakes is assumed in Texas as well as in other parts of the United States. It is certain that the rattlesnakes disappear or become scarce in the vicinity of a new settlement.

Next to the rattlesnake, the water-moccasin (*Trigonocephahts piscivorus Holbr.*) is feared on account of its poisonous bite. This species usually lives near water or moist places. It is easily recognized by its short, stout body, its very broad, but flattened head and the dark color of its body on which a row of dark brown spots appear, running parallel with the backbone.

Very often the chicken snake was brought to me. It often attains a length of five to six feet and although not poisonous, becomes a nuisance to the settlers, since it devours hen eggs and chickens. To catch the latter, it will even climb trees on which the chicks roost at night. One time I retrieved three undamaged eggs from the belly of a chicken snake. The shell of the eggs undoubtedly dissolves after some time through the action of the gastric juices.

A number of small, non-poisonous and often beautifully colored snakes inhabit the prairies and woods. Among them is a graceful and harmless snake of grass-green color (*Leptophis aestivus Holbr.*).

The bullfrog (*Rana pipiens L.*) was also frequently found in the Comal Creek and his voice, indeed resembling the distant bellowing of cattle, was heard especially at night.

The order of birds is also well represented in New Braunfels. As one would expect, most of them are similar to the ones native to other parts of the United States, especially the South. Songbirds are by no means uncommon, although such excellent singers as our nightingale and lark are not found. The mockingbird (*Turdus polyglottus Wilson*) as large as a thrush, with grey plumage is a beloved songster, whose ever-changing song one never tires of hearing. The beautiful red cardinal (*Fringilla cardinalis Nutt.*) has simpler, but more cheerful notes. In addition to those mentioned, there are a host of other small songsters. If it should seem to the German immigrant in America that the woods are not populated with birds, especially songbirds, as in the homeland, this is not the case. In the center of the vast German forests and uncultivated land, comparatively few songbirds would also be found. The abundance of bird-life is usually dependent on the nearness to human dwellings and on the more extensive cultivation of fields.

Among the beautiful and bright colored birds, the bluebird with its bright, sky blue plumage (*Sylvicola coerulea Wils.*) in addition to the cardinal already noted, deserves especial mention. During the summer months one can also see the dainty hummingbird (*Trochilus colubris L.*) flying about at New Braunfels.

In sharp contrast to these dainty inhabitants of the air are the clumsy buzzards (*Cathartes aura L.*) which are very common throughout Texas. They are provided with a keen sense of smell and appear immediately whenever an animal has fallen. Due to their usefulness in disposing of carrion, it is punishable by fine in Louisiana and Texas to kill them.

Of the wild quadrupeds or mammals appearing in the region of New Braunfels, the American deer is the largest, for the buffalo, which were still seen on the heights north of New Braunfels when the first settlers came under Prince Solms, had retreated long ago and did not come within many miles of the city. The deer also disappeared quickly in the immediate vicinity since they were hunted extensively by the settlers and, in addition to this a band of Lepan Indians and other tribes appeared from time to time who waged a veritable war of extermination against them.

During the earlier colonization period, bear (*Ursus Americanus Pallas*) were shot by the colonists in the immediate neighborhood of New Braunfels on the forested banks of the Comal, but they have since retreated entirely.

However, in the summer of 1846, several cougars (*Felis concolor L.*) were shot near the city. The brown lynx (*Felis rufa Temminck*) found throughout America, still lives in the forest along the Comal, and is killed occasionally. The dainty ocelot (*Felis pardalis L.*) also lives in trees and on account of its vividly spotted fur reminds one of the tropical form of the cat family.

Wolves (*Lupus occidentalis Richardson*) may be found everywhere singly on the prairies of New Braunfels. Very common are also the skunks (*Mephitis putorius Gmel.*) widely distributed throughout America, and the opossum (*Didelphis Virginiana L.*) much dreaded as an enemy of the domestic fowl. However, another animal, usually found in all the forests of North America, and also common in eastern Texas, namely the raccoon (*Procyon lotor Harl.*) is seldom found in western Texas and I do not recall seeing a single specimen shot near New Braunfels during my stay.

Mexican peccaries (*Dicotyles torquatus L.*) were encountered in small herds several times on the forested banks of the Comal and several were shot. But they withdrew quickly from the settled regions due to their great fear of man.

Several species of squirrel are also numerous in the forests on the banks of the river; among them the peculiar flying squirrel (*Pteromys volucella Harlan*) which is distributed throughout the United States and which can easily be tamed.

So much for the present about the fauna of the region surrounding New Braunfels. More remarks about certain species will be found in the following pages of this book.

To give a similar summary of the flora would hardly be possible, since it is so extensive. I would but make this general observation here, that, considering the latitude, one is impressed on the first glance with the absence of plants tropical in character. The newly arrived German will find the forests of western Texas similar to those of his native country. He does not find any species of trees which appear foreign to him such as palms and the like. The evergreen trees are only represented by the live oaks.

The herbs also are not so conspicuous as to draw the attention of the European newcomer. Only the yucca and cacti are such strange plants which cannot be compared with any in the native country. Several species of both grow in western Texas. One species of the yucca with large, swordlike stiff leaves reaches a height of from ten to twelve feet and reminds one, in manner of growth, of the palms. Among the cacti are none of the tall, treelike species found in southern Mexico, but the majority of them are the low round species of the genus *Echinocactus*, *Echinocereus* and *Mammillaria*. Only two kind of cacti reach a height of several feet and are shrub like (*Opuntia vulgaris L.* and *O. frutescens Engelm.*). The agave (*Agave Americana L.*) is not found in the region of New Braunfels, but appears in southwest Texas toward the Rio Grande.

The scientifically trained botanist will find that the flora of western Texas differs materially from that of northern Europe. Even when compared with the adjoining Southern States of the Union, such as Louisiana and Arkansas, many peculiar species and kinds appear. The flora of western Texas is like the fauna, forming the connecting link between the flora of the Southern States on both sides of the Mississippi and that of Mexico.

CHAPTER XI
A MEXICAN RANCH—WACOE INDIAN CAMP—
UNSUCCESSFUL TURKEY HUNT—GEOLOGICAL OBSERVATIONS

After I had learned to know the immediate vicinity of New Braunfels fairly well through my many trips, I made excursions to more distant places. On March 10th, I accompanied two agents of the company, Mr. von Coll and Mr. Bene on horseback to a ranch some seventeen miles south of New Braunfels, owned by a Mexican named Flores. This trip was undertaken to ascertain whether a number of oxen, belonging to the Verein, and which had been missed for some time had not perhaps joined the cattle of the afore named Mexican. Our journey to this ranch first led us over an open prairie which was already covered with verdant green and the countless blossoms of an onion-like plant. Later we came to mesquite trees which were just unfolding their tender, feathered, succulent green leaves. Soon we neared the fringe of trees bordering the Guadalupe, where other groups of forest trees appeared, particularly the elm with its rose-like leaves. These trees were everywhere adorned in the richest green, and with a beautiful lawn beneath them presented a pleasing park-like view. Here and there the huge, white clusters of flowers of the yucca, more than a foot long, rose above the shrubs like a bouquet of lilies.

Before reaching the ranch, we passed two American farms which were just being established. The families, only recently arrived, were living in huts, erected for temporary use. The field in which the much needed corn was to be raised during the summer, was not enclosed with a regular fence, but the colonists had contented themselves with surrounding it with a brush fence, made of felled trees and brush, as a protection against the cattle roaming about.

The Rancho Flores presented an altogether different aspect than did the farms of the American colonists. The spacious yard was enclosed according to Mexican custom with a palisade of mesquite trees rammed into the ground, which were further connected with strips of raw oxhides. Several clumsy Mexican carts with disk-like solid wooden wheels stood near the entrance. A little farther on, sheep and goats were roaming about, a rather strange sight in Texas. The one-story home and the various out-houses were made partly of logs lying horizontally over each other, and partly of logs standing perpendicular, the crevices of which were filled with clay. These were built securely in order to successfully withstand an Indian attack, which in times prior to the founding of New Braunfels, was not improbable, since there was no settlement north of the Guadalupe.

Behind the buildings is a large fenced-in field. On the other side close to the ranch flows the Guadalupe, forming here a rushing little waterfall. The river with its abundant, crystal clear, rapidly flowing water, shaded by beautiful cottonwood and other trees of the bottoms, presents a most pleasing view. The proximity of the river, the excellent farming land, and the unlimited pasturage with its tender grass on which thousands of cattle could find an abundance of nourishment, are natural advantages, which combine to make this ranch a valuable piece of property.

We did not find the owner at home but were received by his wife, who welcomed us in a friendly but dignified manner. She was an elderly lady with unmistakable Spanish features, and was rather corpulent, which is characteristic of Mexican women, while the women of the American settlers usually become lanker with increasing age. After a little while we were served a meal, consisting of chicken ragout, liberally seasoned with Spanish chile-pepper, baked eggs, cornbread and excellent coffee. The Señora Flores served us our coffee with great dignity. She spoke only Spanish and since neither of us knew much of the Spanish language, the conversation was naturally limited.

When my companions had received the desired information with reference to the missing cattle, from the Mexicans working for Mr. Flores, we returned after the hottest hours of the day had passed, arriving at New Braunfels late at night. The glorious spring weather and the changeable beauty of the landscapes, almost untouched by the hand of man, made this trip one of the most pleasant I can recall during my stay in Texas. A number of other trips were made owing to the circumstance that the Verein had engaged a number of young men, organized along military lines, who were to serve as an armed escort to the long contemplated movement of colonists to Fredericksburg. In order to acquaint these young men with camping outdoors and in general to accustom them to a life in the wilderness, the leader of this small band of mounted men, Mr. B. (formerly lieutenant in the Prussian army) made a number of excursions, lasting several days each. I was glad to join them on these trips, since it enabled me to observe and collect at certain places with perfect safety, where without companions it would have been inadvisable to go, due to a possible Indian attack.

The destination of our first excursion was a point lying about eight miles north of New Braunfels on the Guadalupe, where at one time the Wacoe Indians had camped. To reach this place we first ascended the cedar-covered hill slope north of New Braunfels. We rode past Mission Hill, mentioned previously, and then crossed several rocky ravines, usually dry, which are so characteristic of the stony hill country north of New Braunfels.

For miles the eye roved over the grass-covered, almost treeless heights and to such an extent did nature present a picture of original primitiveness undisturbed by human activity, that one could imagine one's self removed hundreds of miles from civilization. The wilderness does indeed extend uninterruptedly toward the northwest to the upper reaches of the Rio Grande in New Mexico. We saw very few living creatures on these heights. Only occasionally one could see a small herd of deer in the distance.

After a ride of one and one-half hours, a small tributary valley led us down to the Guadalupe, where close to its forested bottom lay the abandoned Indian camp on a hill among scattered live oak trees. The bent twigs, resembling a bower, which formed the huts or wigwams after being covered with skins, were still standing, and in the middle of each, the little fireplace was still plainly visible. A more pleasant and appropriate site for a camp could not have been chosen. The Guadalupe, which skirts the foot of the hills, forms here a noisy waterfall, and below it the river enlarges itself into a deep basin of crystal clear water. Just above the water fall stands a row of mighty cypress trees (*Taxodium distichum Rich.*) diagonally across the stream, so that the bases of the trunks are surrounded by the rapidly flowing water, causing one to wonder how the trees could have taken root at such a place. Like the mesquite trees whose leaves appear rather late, the cypress were just now unfurling their dainty, succulent green foliage. A short distance above the fall is a ford, the only one above New Braunfels where the river can be crossed conveniently on horseback.

The proximity of this ford explains the presence of an Indian camp at this particular spot. We also chose the same camp formerly occupied by the Indians. A fire was kindled to cook coffee, which is always the first and indispensable refreshment in the Texas wilderness. A few hunters were then sent out to kill some game if possible. One of these, a brave young man and zealous hunter, formerly from Montabaur near Coblenz, who had bidden his home farewell in order to satisfy his desire for hunting in Texas, was extremely fortunate on this day, for in a short time he had bagged three deer and a turkey. We therefore had an overabundance of meat and the choicest pieces were quickly prepared in various ways.

When twilight descended, the voice of the whippoorwill (*Caprimulgus vociferus Wilson*) known to all inhabitants of North America, resounded from the neighboring bushes, resembling somewhat the call of our quail, and from the tops of the high trees along the river's bank, the hollow gobble of the turkeys could be heard. We decided to bag some of the latter on the following morning at daybreak, for at this time they are less wary and betray their presence regularly through their gobbling with which they call the hens.

Accordingly we retired early, our beds consisting of woolen saddle blankets spread out under the live oak trees. The beautiful warm night such as is seldom to be found in Germany in July obviated the necessity of further covering and all fear of rain was dispelled when looking at the beautiful, deep blue, starlit sky. We arose before break of day and sneaked quietly to the place where the turkeys had been heard. As soon as day dawned, we could indeed hear the gobbling from several directions. Despite this, our hunt was a failure, for the steep, rocky bank prevented us from approaching the high trees in which the turkeys were roosting. Moreover, when one member of our party discharged his gun, mistaking a bundle of Spanish moss hanging in the treetop for a turkey, the real turkeys flew to the trees on the other bank of the Guadalupe.

The rest of the day was utilized in making excursions in the neighborhood of our camp. At one time we followed the course of the Guadalupe upstream. A narrow but luxuriant primeval forest enclosed its bottom. Some distance above the waterfall we came to a spring, fringed with dainty bushes of the palmetto and shaded by beautiful forest trees. It gushed forth with a volume and intensity equal to the springs of the Comal. Still farther we followed the tracks of a wagon which was the only indication that this wilderness had been trodden by civilized man. The tracks led us to a place where several enterprising Americans had made shingles of cypress wood. In apparent danger of being killed by Indians but at the same time assured of considerable profit, they had remained several weeks here. This reflects the typical American spirit of not being deterred by dangers and difficulties.

We were not able to penetrate much farther than this point, for here the valley of the Guadalupe became a steep, rocky gorge. Near us lay a hill, several hundred feet high, covered sparsely with stunted cedars. On its almost perpendicular slopes, rising directly from the water's edge, layers of a pale yellow limestone of the cretaceous formation were visible everywhere on the surface.

If someone, acquainted with only the inhabited portions of Texas with its undulating hills and extensive lowlands along the coast, where a single piece of gravel can scarcely be found, would suddenly be transported to this region, such a person would firmly believe that he were in some other country. To such an extent does the hilly country, composed primarily of firm limestone layers of the cretaceous formation, differ from the other parts of Texas.

Our camp and its immediate surroundings was particularly adapted for my purposes. The composition of the soil could be readily observed, due to the deeply cut gullies and ravines, and I had the opportunity to gather numerous fossils. In the small valley in which the Wacoe camp lay, thousands of specimen of the peculiar *Exogyra arietina* were found on the surface; however, the *Gryphaea Pitcheri Morton* were less plentiful and the *Pecten quadricostatus Sow.* were rare. Quite close to this camp, I observed fragments of large ammonites and *Ostrea carinata Lam.*

After spending several days in the wilderness in this manner which to me proved very attractive, we returned to New Braunfels just in time. The spring weather up to this time had been most pleasant with the exception of a severe thunderstorm. For weeks there had been no sign of the piercing northers, but in the evening upon our return, after quite sultry weather, a severe norther accompanied by cold rain, swooped down suddenly, and even on the following day, Easter morning, it continued to blow. One could scarcely find protection from its penetrating blast at the fireplace of the flimsily constructed houses.

On the 15th of April, a similar trip was undertaken by the same armed company to the Cibolo River, fifteen miles distant, lying halfway between New Braunfels and San Antonio. Several miles this side of the Cibolo, we saw in the distance, four mounted men coming toward us, who halted abruptly when they perceived us and who apparently were undecided whether to advance or to retreat. However, they finally came toward us and we had an opportunity to inspect them more closely to determine more definitely who they were. The two men riding in advance on beautiful bay horses were evidently not native Americans. A certain less businesslike, but more elegant appearance than is usually found among the American travelers in this region, and the further fact that instead of the long American rifle hung across the shoulder, they carried a short German rifle slung across the saddle and a few other characteristics indicated the European origin of both riders. The other two men were evidently servants of the former. After exchanging greetings we were informed that the first two gentlemen were Mr. Carl Obermeyer from Augsburg and a Pole, Count Olinsky.

The two gentlemen, engaged in a pleasure tour of Texas, seemed just as pleased to find friendly Europeans instead of marauding Indians, for which they had mistaken us from the distance, as we were surprised on our part to find European tourists in this unknown wilderness. After the gentlemen had given us the promise to spend a few days in New Braunfels, so that we would still find them upon our return, we all continued our journey.

The bed of the Cibolo was dry as usual. Only in several hollows below the road to San Antonio was water found. We selected our camping place at one of these hollows, since it contained drinkable, cool water, due to the high cottonwood and poplar trees which shaded it. Not far from this spot, a band of Indians, likely Tonkoways, had recently camped, for at several places large heaps of deer hair, a foot high, could be seen, which were left after the hides had been prepared for the manufacture of clothes. These piles of hair were also a mute evidence of the extraordinary war of extermination, which the Indians wage among game animals.

Our horses found a plentiful supply of food on the mesquite-covered level plain, which extended on both sides of the river. The only thing missing to make the valley of the

Cibolo suitable for settlements is constantly flowing water. Wells will no doubt supply enough water in some parts of Texas for use in the home, but they could not satisfy the needs for cattle raising on a large scale.

When we first neared the waterhole, we saw two birds fly away, one a large heron and the other an anhinga (*Plotus anhinga Lin.*), an aquatic bird with a thin, long neck, resembling a snake, and a long, pointed bill. This bird is not uncommon in Texas as well as in other Southern States.

A specimen of a silver fox (*Canis Virginianus Gmel.*) was seen among the rocks in the bed of the Cibolo. It is not uncommon in western Texas. A species of wild grape, the only one in Texas which bears white grapes and which is perhaps best suited for cultivation, grew among the limestone rubble of the riverbed, and had just unfolded sweet smelling tufts of blossoms.

As far as the banks of the Cibolo are rocky, they are lined by cedar bushes (*Juniperus Virginiana Lin.*). The cedars of western Texas do not extend anywhere else to the fertile soil of the plains, except in a few bottoms. There they form high, slender trunks on which the lower limbs appear to have died, leaving a sparsely foliated crown. Among the cedar bushes grew yuccas and cacti.

Also from a geological standpoint our camping ground was not without interest. Immediately above this point where the road from New Braunfels to San Antonio de Bexar crosses the Cibolo, white horizontal, deposits of marly limestone of the cretaceous formation form peculiarly shaped cliffs, ten to twelve feet high, apparently formed through the erosive force of the water. These cliffs contain numerous fossils. Very common is the smooth *Exogyra B. laeviuscula* as large as two inches. In addition to these, fragments of Ammonites, not described, also Baculites, Spondylus and Micraster were observed by me.

Chapter XII

Immigrant Train—Letter from Gov. Henderson—
Return to San Antonio—Indian Horse Theft

In the days following our return, the thoughts of the
people of New Braunfels were primarily occupied with
preparations for the departure of a company of emigrants who intended to colonize on the Pedernales River.
Already in the fall of the previous year an expedition under
von Meusebach had chartered the location for the colony
on the Pedernales. It had also bought the land and had even
made an attempt to establish the colony. The latter attempt,
however, was soon abandoned, since at that time it was impossible to provision the colony regularly.

Now the colonization was to be undertaken with greater
energy and on a more extensive scale. The gradual accumulation of immigrants arriving from the coast who had to be
maintained by the Verein in New Braunfels until land could
be assigned to them, made it urgent upon the management
of the Verein not to delay any further in carrying out this plan.

A number of Mexicans from the region near San Antonio had been engaged with their ox-carts to transport the
baggage and the necessary supplies for the colonists. But it
became known several days before the date set for the departure that the Mexicans would not appear despite their
promise, very likely because on a second thought, the whole
journey seemed too dangerous, owing to the Indians. The
Verein under the direction of Herr von Coll who served
as manager during the absence of von Meusebach, had to
equip several of their own wagons for this purpose. Lieut.
Bene was appointed leader of the train.

The departure finally took place on April 23. The train
consisted of 16 wagons, drawn by two or three yoke of oxen,
and 180 persons, including the mounted convoy, which accompanied the expedition for their protection.

We were filled with a feeling of heartfelt sympathy for the
many people who were departing for a destination wholly

unknown to them and who, in all probability, were to face a difficult and dangerous future.

The departure of this company proved of general interest also, for through it the first settlement of civilized human beings was to be founded in the northwest hill country of Texas. Up to this time this country was only sparsely and exclusively inhabited by roving bands of Indians; and whereas the Anglo-American was always the first to advance into the western wilderness, here the German was to be assigned this role. It was therefore with a feeling of sympathy that I followed the long train as it skirted the woods of the Comal Creek until it finally disappeared over a hillock in the prairie.

For several days many signs indicated that the warmer season was approaching. Almost every day we had a thunderstorm accompanied by a torrential rain, which, much like tropical rains, precipitated an unusually large amount of water in a short time. This happened usually at night and on the following morning the sky was once again cloudless and clear. The air was continually warm, often sultry.

Nearly every day someone brought me a brown spotted chicken snake, often five feet in length. Rattlesnakes were encountered frequently in the prairie. Green lizards, a span long, ran across our path in great numbers as quick as lightning and daily one could see in the basin of the Comal creek, alligators sunning themselves, lying motionless near the surface of the water, the snout only projecting. The dainty little hummingbird also whizzed by frequently; and the cream-colored flycatcher (*Milvulus forficatus Gmel.*) flew about noisily, alternately opening and closing its long forked tail.

At night the cicadas, a kind of small cricket, named "Katydid" by the Americans, entertained us, as did treefrogs with their peculiar, continuous monotonous noise, which gives the beautiful Texas nights their distinctive character and which is also noticed in the Northern States, as for example in New York, but in a lesser degree.

The flora also announced the progress of the season. The mesquite trees were covered with small yellow blossoms. In the cornfields near the buildings of the Verein, the succulent green blades of the corn were already a foot and a half high.

The flowers of the onion-like plants had disappeared in the prairie and in their stead a manifold variety of blooming plants took their place. Among them, two species of *Euchroma* distinguished themselves through their vermilion flowers. The small, stemless yucca with small linear leaves (*Yucca filamentosa L.*) had sent up its two-foot high scape covered with white bells. The various kinds of cacti also began to unfurl their vividly colored blooms. In the common variety they were sulphur yellow with red base, in the foot-high semiglobular *Echinocactus Texensis Hpfr.* a beautiful rose red.

On May 4, the management of the Verein received a letter from the Governor of the State, Colonel Henderson, in which he advised against carrying out the plan of sending an expedition to the Pedernales of which plan he had heard, declaring that he had reason to believe that the Comanches would prove hostile toward the settlers, and that the intended military protection on the part of the government could not be granted, as all available troops were needed in the war against Mexico. This letter caused great consternation in New Braunfels, for it left the impression that the colonists at the Pedernales were threatened by great danger, or had already met with misfortune.

Through a bell hanging on a pole in the market square, the inhabitants of the town were called together for a meeting, in which it was to be decided in what manner help could be sent to settlers on the Pedernales. It was decided to send a company of mounted volunteers to Fredericksburg, who should inform the colonists of the letter from the governor. In case the settlers thought it best to return to New Braunfels, the former should act as an escort.

With commendable alacrity twelve young men volunteered to make this dangerous trip. All armed themselves to the limit. For example, the leader whom the band had chosen, had eighteen shots in readiness for the Indians, since he carried two Colt pistols with revolving cylinders, each containing five shots, and a rifle manufactured on the same principle, containing eight shots. Armed in this manner, they left in a cheerful mood on the next morning.

On May 7, business led one of the officers of the Verein to San Antonio. The weather being ideal, I accompanied him since I wished to view the town in its summer garb. This excursion was one of the most pleasant ones I recall during my entire stay in Texas. Never had the prairie appeared to me more as a charming natural garden or park on a large scale. Countless blossoms—many of which had only recently been introduced in our gardens as ornamental plants, such as *Gaillardia picta* and several species of *Coreopsis*—formed natural flower gardens miles in extent. The luxuriant growth of vegetation was especially noticeable at the spring of the San Antonio River. This time we followed the road below the springs which compelled us to cross the river near the city. The deep blue of the incomparable beautiful springs was contrasted most charmingly by the succulent green of the trees and shrubs which enclosed them. The cacti, growing in great abundance on both sides of the road were covered with sulphur-yellow blooms to such an extent that each one of the large fleshy leaves formed a garland of flowers. The mesquite trees, eight to ten feet high, with which the entire plain in the vicinity of San Antonio was covered, were resplendent in the beautiful garb of dainty green leaves.

In the city itself we found great excitement among the American population. Nearly everyone was busy equipping himself to enter the war against Mexico as volunteer. Already for several days bad news had arrived from the Rio Grande, according to which the American army had suffered severe casualties and General Taylor was reported to have been surrounded by a Mexican army. Many Mexican inhabitants had left the city and had crossed to the other side of the Rio Grande in order not to live in the land of the enemy of their nation during the war between their compatriots and the hated Yankees.

We visited the colorful regiment of dragoons of the American Army stationed near San Antonio, commanded by Colonel H. Harney, from whom we asked military protection for New Braunfels against possible Indian attacks, which were likely to occur during the course of the war. He was a man of middle age, and despite the simplicity of his blue uniform, presented a stricter military appearance than one is accustomed to see among the officers of the American Army. His erect carriage and the serious expression of his manly countenance betrayed at the same time determination and inflexible willpower, through which he distinguished himself later on several occasions during the course of the Mexican War. He gave us a friendly reception and assured us, that he intended to station a company of mounted riflemen at New Braunfels for the protection of the colonists.

In the gardens of the city the pomegranate and fig trees were resplendent in their dark green foliage. As regards the life of the city, it appeared the same as during my first visit. The Mexican women still bathed unabashed in the river and the fandangos took place regularly every night. However, one saw fewer dignified señores wrapped in gaudy blankets, on the streets.

In the cool of the evening we rode out to the nearest mission, in the vicinity of which we met a squadron of dragoons in a romantic camping ground. The white tents were erected in two long rows with an alley between them in which the horses were tied to protect them if possible from the daring marauding Indian bands.

How much the latter were to be feared, we had occasion to find out immediately upon returning to New Braunfels. We were met with the news that, on the day previous just before sunset, twenty horses had been stolen by the Indians from the pasture near the city. The theft was carried out with great daring almost within view of the inhabitants of the city and with such rapidity that, although it was discovered almost immediately, successful pursuit of the thieves could not be thought of. Two of the horses driven off returned the next day with Indian arrows sticking in their paunches. Very likely they refused to be driven as willingly as the others and in their anger the thieves had tried to kill them.

This was the first important reverse which the settlers suffered from the Indians. However, several weeks prior to this, a pair of oxen belonging to a settler, had been found, killed with arrows, on the cedar forest of the hill slope, hardly two miles from the city.

With reference to the horse theft, I should like to remark, however, that some of the colonists blamed it upon some Mexicans near San Antonio who wanted to supply themselves with horses before crossing the Rio Grande. As a matter of fact, it is a general opinion in West Texas that the Mexicans have the same inclination and skill for stealing horses as the Indians. As a rule the Indians choose the time just prior to sundown for their horse thefts, because they then prevent an immediate pursuit since the darkness falls rapidly in the southern latitude. If the pursuit is not undertaken immediately, it is usually without result, for during the night the thieves gain such a lead that it is difficult to overtake them even with the fleetest horses. Furthermore, they usually flee to the hills where on the stony paths a rapid pursuit is almost impossible. Should the pursuers nevertheless come near them, perhaps after several days, they protect themselves by scattering in all directions with their prey. Then the trail which the pursuers followed until now suddenly becomes indistinct and, disheartened, they abandon further pursuit.

On the day of our return to New Braunfels the mounted company of volunteers also returned from Fredericksburg. They had found the colonists busily engaged in the building of their houses. With reference to the letter from Governor Henderson, they had decided to remain and to depend upon their own means of protection. Indians had so far not been seen, except a band of friendly Lepans who were on a hunting trip.

For some time individual wagons had arrived from Indian Point with immigrants. Most of these were sick and especially those, who due to dysentery had been reduced to a mere shadow of their former self, presented a most pitiable sight. They also brought the sad news that of the German immigrants camping at Indian Point, one hundred forty had already died.

In marked contrast to this virulent disease was the beauty of the weather. Since the beginning of May it was almost uninterruptedly clear and rainless. The temperature varied during the day between 79°F. and 84° F. It is hard for the person from the north to understand, why the cloudless clear sky and the warmth of the sun, so often missed in his native land, should be accused of causing sickness. A longer stay, however, will soon convince him, that the summer months throughout the Southern States of the Union are the real time of harvest for the physician, while during the winter months his service is scarcely needed.

Chapter XIII
The Plantation "Nassau"—La Grange & Rutersville—
Austin & The Capitol—Germans Murdered by Indians

For some time I had cherished the wish to become acquainted with the more distant parts of the country through personal observation. An opportunity to gratify this desire presented itself as I found an agreeable traveling companion in the person of my esteemed friend Herr W. Langenheim from Braunschweig. We decided to visit Herr von Meusebach at the plantation Nassau near La Grange, which was owned by the Verein, and to return by way of Austin. As far as Nassau, we were accompanied also by Mr. Henry Hoffmann, a well educated young man from near Danzig, who had come to Texas to engage in agriculture. He had the intention of looking up Herr von Meusebach through whom he hoped to effect the return of several hundred dollars which he had deposited with the Verein, and which he was unable to obtain at New Braunfels, owing to the shortage of funds.

We left New Braunfels on May 18, mounted and well-equipped to camp out of doors occasionally although our journey in general led us through well-settled regions. It is often more desirable to camp out of doors during the beautiful Texas summer nights than to make use of the American farmhouses where the comforts do not meet the requirements of our German taste.

The first part of the way along the Guadalupe was known to me from my journey here from Houston, but it had an altogether different appearance in its summer garb. At a point several miles beyond Gonzales and Peach Creek where the road to Columbus and La Grange branches off from the main road, the country was new to me.

The greater part of the journey between Gonzales and La Grange on the Colorado led through a sandy, hilly country, covered almost entirely with post oak forests in which there were only isolated settlements. Especially monotonous and desolate is the region on top of a plateau which appeared to be a continuation of the Big Hills which we crossed on our way from Columbus to Gonzales. Not until about twelve miles from the Colorado does the forest end and one enters a pleasant, gently undulating country, composed mostly of open prairie with scattered groups of live oaks and other trees. The soil also is much improved here, as the presence of numerous farms indicates. The subsoil is composed of thin, horizontal slabs of less firm, chalky sandstone.

Upon nearing the Colorado, we saw rising to the south a long hill, three hundred to four hundred feet high, the so-called Buckner Heights, whose north and also east side facing the river, had such steep slopes as are seldom found in the gently rounded hills of Central Texas. Here the Colorado has only a small forested bottom and is about as large as our Weser in its middle course.

After we had ferried across the river, we were but a short distance from La Grange, which is situated pleasantly in a bend of the river and which presented a cheerful appearance with its white painted frame buildings. The distance from La Grange to New Braunfels is about one hundred ten miles which we covered in almost three days.

Our destination was only twenty-five miles from this place. The road first led us as far as Rutersville through sandy, almost level land, overgrown with oak trees. The latter named place is a settlement founded by the Methodists, who also maintain an academy here. According to a common North American custom, the school building lies isolated on a hill and presents a stately appearance from the distance.

Since we had left Rutersville several hours before sunset, we hoped to reach the plantation Nassau that day as it was only thirteen miles distant. However, we were destined to be disappointed, for shortly after the darkness had fallen, and when we were only a few miles distant from the plantation, according to our reckoning, we lost our way and after wandering around several hours in the meadows and among the bushes, our own fatigue and still more that of our horses, finally compelled us to stop. We unsaddled our horses, tethered them to a live oak with a long rope carried for this purpose by each traveler to afford them plenty of room for grazing in the long grass, in which we also lay down, missing shelter much less under the beautiful starry sky and balmy night air than a hearty meal which we had hoped to find at the plantation after the long ride. But also this latter inconvenience was forgotten as sleep soon overtook us.

While wandering about, we were entertained by a spectacle, the like of which surpassed anything in vividness I remember seeing in Texas. Glowworms flew over the long grass of the prairie in such countless numbers and emitted such bright rays of light that the eye was actually blinded or dazzled and one imagined one's self to be in the midst of a rain of fire.

Early the following morning we were awakened by the crowing of the cocks, and when we had rubbed the sleep out of our eyes, and looked about us, there, as if mocking us, lay the manor house of the plantation on a hill, scarcely a gunshot distant.

We surprised the entire household, who were still sleeping, not in the house, but on the gallery surrounding it. A nourishing breakfast composed of coffee, ham, eggs, and cornbread, served by a barefooted little negro girl with the romantic name "Amanda," soon made us forget the privations of the previous day.

During the course of the day we took occasion to inspect the location and the arrangement of the plantation a little closer. The plantation Nassau was founded in the year 1843 by Count von Boos-Waldeck, who in company with Prince Leiningen remained in Texas for some time investigating conditions pertaining to German colonization for persons who later organized the Mainzer Verein. It contained 4,440 acres of land, a so-called league or square league. The greater part of it is open prairie with but a smaller portion forested. Excellent oak logs for building and for fences are found in great quantities and the plantation is especially rich in cedar wood which is so valuable for many purposes.

The manor house lies on a hill covered with oak trees and is separated from the other farm buildings. It is one of the best constructed and most comfortable houses I have seen in Texas. Nevertheless it is only a log house, but of course not the kind the American back woodsman first erects as a shelter when he penetrates the wilderness.

The whole house is built of rough-hewn oak logs carefully grooved, lying horizontally over each other. It is separated into two parts, according to the custom of the country, forming in the center an open, covered passage, which offers the inhabitants a cool, pleasant resort in summer. The two longer sides of the house face north and south, so that the prevailing south winds in summer can circulate freely through the hall. On these two sides, the roof projects about ten feet and is supported by wooden pillars forming the aforementioned galleries, whose floors are two feet above the ground. On each end of the house is a fireplace built of ashlar stones reaching several feet above the top of the house which gives to the whole building a stately appearance. Most of the farmhouses of Texas have fireplaces built of thin logs which are placed over each other to form a square. The cracks are filled with mud or clay.

Next to the manor house stands an ordinary log house, occupied by a negro family, the servants of the household. The head of this family was a very valuable negro, who among other accomplishments thoroughly understood the trade of blacksmith and who could easily have earned three dollars a day at this trade. Several times he could have been sold for two thousand dollars. Since he always conducted himself properly, he was treated with a certain consideration and he boasted that during thirty years he had not received a single beating from his master. A German overseer of the plantation, inexperienced in the handling of negroes, decided to whip him on account of a supposed disobedience, just as was done to the ordinary negro working in the field, but he opposed this and ran away, and could only be induced to return on the assurance that he would not be punished. Nor need one wonder that the consciousness of human dignity awakens in such negroes who, in accomplishments and cultural advancement, are almost equal to the whites.

With reference to the case just mentioned I should like to remark in passing that the German seldom understands how to handle a large number of negroes. It requires a combination of great determination and a certain amount of indulgence in order to strike a happy medium and one must have an intimate knowledge of the disposition of the negro. The farm buildings belonging to the plantation lie about a gunshot distant from the manor house at the foot of the hill. There are barns, storage houses, negro cabins and a house for the overseer. All are rough log houses made of roughly hewn logs, covered with shingles, which, like most buildings of this kind on an American plantation have no particular pleasing appearance, and in neatness and substantial construction do not compare with the farm buildings of a large German estate.

There were only nineteen negro men and women on the plantation which is a comparatively small number, for on the sugar and cotton plantations of the Brazos there are frequently fifty, and on the large sugar plantations in Louisiana the number often reaches one hundred or more.

Four hundred twenty acres of the 4,400 were fenced in and under cultivation. In previous years cotton was especially planted, but this year it was deemed more profitable to raise corn, tobacco and potatoes. The net proceeds of this plantation were never large which is readily explained by the fact that none of the owners remained permanently here and the care of everything was therefore left to an overseer or manager, among whom there were also frequent changes.

The region in which the plantation Nassau lies is similar to the entire strip between La Grange and San Felipe, a rather characterless hilly country, in which open prairies and oak forests alternate. Several small, sluggish streams wind through it. The peculiar flora of the western and southern Texas, the yuccas, cacti and mesquite trees are absent here, and nothing reminds one that this is a semi-tropical region. Transported here suddenly one could imagine one's self to be in some parts of Germany, as for example in a certain region of the Rhine. In beautiful landscapes and pleasing appearance this strip of land is far inferior to the regions of western Texas. In the latter region the pasturage is also better, for even if there is an abundance of grass here in the prairies it is not the tender and nourishing mesquite grass of the prairies between the Guadalupe and the San Antonio rivers. This region on the other hand, has the advantage of being nearer the seacoast and therefore has easy communication with the marts of the country as well as a greater wealth of good building material which is not found in the western part of the State.

Numerous German settlements are found in this region, particularly on Cummins and Mill creeks. The hamlet Industry, founded more than twelve years ago by a German, Mr. Ernst from Oldenburg, lies only about six miles from Nassau. Some of the Germans living here have become financially independent through their persevering industry and are well satisfied with their lot in every respect.

Herr von Meusebach had been living at the plantation Nassau for several weeks under painful and unpleasant circumstances. After making all the necessary arrangements in January in Galveston for the care and transportation of a great number of immigrants, he was obliged to leave there in order to hide from the many creditors of the Verein, whom he was unable to satisfy, since no new remittances from Europe had arrived for several weeks. He went to the plantation Nassau to await the arrival of more money. He did not venture to go to New Braunfels or Indian Point for at these places the dissatisfaction due to the shortage of funds in the treasury of the Verein had reached such a point that his appearance there without sufficient funds to carry out the agreement entered upon with the Verein would likely have resulted in excesses against his person. With his peculiar business ability he was in the meantime constantly engaged in raising small sums of money to meet the most urgent needs and to keep the persistent creditors from taking legal proceedings against the Verein.

After spending a few days on this plantation, we continued our journey on June 1 to Bastrop and Austin. We first returned by way of Rutersville to La Grange where Mr. Hoffmann, who took the shortest route home to New Braunfels, left us. The road from La Grange to Bastrop, but forty miles distant, follows the left bank of the Colorado, leading over sandy, infertile pine-covered hills and through the rich level bottoms of the river. Just opposite La Grange, the peculiar sterile, silicious soil appeared in which in addition to red rubble stones, larger and smaller pieces of silicified wood occurred very frequently. Only post oak (*Quercus obtusiloba Mx.*) and yellow pine (*Pinus taeda L.*) grow on the long stretches of gravel and sand of central Texas. Both species of trees have the peculiarity of excluding all other trees where they grow whereas other deciduous forests of North America are composed of a great variety of trees.

In distinct contrast to the prevailing uniformity of the vegetation and sterility of the soil, stood the forested bottoms of several brooks, flowing into the Colorado, through which our way led. In this deep, black soil, the bottom vegetation grew as luxuriantly as is the case along the larger rivers such as the Brazos. The huge poplar trees on which the grapes and other vines climbed, were again conspicuous.

As no dwelling was near when darkness descended, we camped out of doors in the clearing of a pine forest, where we found near us good food for our horses and drinkable water for cooking coffee. Under a beautiful starry sky and balmy night air we sat and chatted for several hours before retiring. My companion told me many interesting things about his life in America, a life very eventful as well as colorful. Since it is so similar to those of countless numbers of Germans of the upper class who immigrated to America, a

short sketch of his life is in place here. The reliable character
of the narrator, not given to boasting, vouches for the truthfulness of his statements.

Mr. Langenheim came to America in 1830 after having
followed the profession of jurist several years in his native
town of Braunschweig. Shortly after his arrival in New
York, he joined a company of emigrants composed mostly
of Irish and a few Germans, who intended to found a colony
in southern Texas on Aransas Bay. The colonization was
successful and Wm. Langenheim was just beginning to enjoy the comforts of his new home, wrought by the work of
his hand, when the War between Mexico and Texas which
gained Texas its independence, broke out, calling every
able-bodied man to the colors. He left his hearth and home
and joined the Texas army in defense of his new country. At
the siege of San Antonio de Bexar by the Texans he distinguished himself by his bravery and resoluteness, working
the only cannon which the Texans possessed with undaunted personal courage.

In the following year while watching a number of horses
belonging to the Texas government, at the lower course of
the San Antonio River, he and a few other Texans were attacked suddenly by a detachment of Mexican cavalry and
were captured. They were dragged several hundred miles
through the wilderness between Corpus Christi and the Rio
Grande, suffering untold hardships, and taken to Matamoros. Here he was incarcerated for ten months, during which
time he learned to know the brutal manner with which the
Mexicans, to their shame, treat their prisoners of war. Twice
he was condemned to death by order of Santa Anna, but
was saved through the noble intervention of magnanimous
Mexican women of Matamoros, who raised a voluntary contribution of several thousand Mexican dollars with which
they bribed the commanding officers not to carry out the
death sentence. After the defeat of Santa Anna at the Battle of
San Jacinto, he and his companions were liberated and em-
barked immediately for New Orleans, since he had no desire

to return to Texas, for his farm buildings had been burned to the ground by the Mexicans at the beginning of the war.

He arrived at New Orleans without any funds just at a time when the paralyzing influence of the well-known disastrous crisis of the year 1837, was felt most in the commercial and industrial world. Bereft of all opportunity of finding employment, he was finally successful in obtaining the necessary means to go to St. Louis. But conditions were no better there and for want of a better occupation, he considered himself fortunate to find a position on the quartermaster staff of a regiment of dragoons of the United States army, just leaving for Florida, for the purpose of subjugating the Seminole Indians. After making the long trip from St. Louis to Florida overland, he took part, for a year and a half, in all the hardships and vicissitudes of the long, tiresome, but inglorious campaign. When a virulent fever, for which this country was notoriously known, brought him to the point of death, he decided to go to the Northern States in order to regain his health.

Not long after this we find him editing a German paper in Philadelphia which occupation he followed for some time. Later he decided to enter an altogether new field. In partnership with his brother who had followed him from Europe, he organized a daguerreotype establishment, which, on account of its manufacture of most perfect portraits in which a new process was employed, established a reputation and is still recognized as one of the outstanding establishments of its kind in the United States. In the spring of of 1846, after a ten years' absence, Mr. Langenheim came back to Texas, whose green prairies and clear sky he could not forget, as is the case with the majority of people who have become acquainted with them. His intention was to make a second attempt to establish his home here, and at this time he was traveling about greeting old friends and selecting a site for his new permanent home. The latter intention he did not carry out however, as family affairs compelled him to return to Philadelphia in which vicinity he now lives.

Should these lines come to his attention, may he see in them a sign of my friendly and grateful remembrance of the pleasant hours spent with him in Texas. On the following day our road led us again over sandy hills and through fruitful prairies, two to three miles long, which lie in the basin of the Colorado. Extensive corn fields of unusual richness, containing more than one hundred acres each, were seen in the prairies. The buildings belonging to the plantations lay mostly on the skirts of the forests surrounding the prairies. Although the large cotton gins standing near most of the plantations would indicate that cotton was chiefly raised, still we did not see a single cotton field until we came to Bastrop.

The extensive immigration of the past year into the western part of the State and the founding of numerous German settlements in the upper Guadalupe Valley had advanced the price of corn, the most important food product, to such an extent that its production promised much greater returns than the raising of cotton, especially since there had been several partial crop failures during the past few years, brought about by worms. In consequence of this the settlers had converted their cotton fields into corn fields.

We also passed several oat fields which were already partially harvested on June 2. This grain is raised quite extensively in the region of Bastrop. Although corn is generally used for horse feed, oats are preferred for the racehorses.

Since we tarried at a number of plantations, we did not reach Bastrop until late at night, although we covered only thirty miles that day. Despite the lateness of the hour we not only found good lodging in the hotel, but also had a good supper composed of coffee, cornbread, fried chicken and canned fruits.

On the following morning we inspected the city, but soon convinced ourselves that like other Central Texas towns it offered little of note. About eighty to ninety frame houses, painted white, stood on several broad, straight unpaved streets. Among them were six to eight stores and three or four saloons. Most of the houses looked rather dilapidated and the place needed a "new start" apparently (as the Americans say) to keep it from losing its appearance of a town altogether. The location of the town was not badly chosen, as it was situated on a fruitful, small plain on which mesquite trees grew, and near the fifty to sixty feet high steep banks of the Colorado. The city was named after a German, Baron von Bastrop, who had planned an extensive German colonization here at the time when the first American settlements were still under Mexican rule. He bought a large tract of land for this purpose, but death prevented him from carrying out his plans.

Prior to the founding of Austin the city was for a long time the most northern settlement in the Colorado Valley and as such was constantly harassed by the Indians. It was abandoned several times for this reason. In the course of the day we continued our journey to Austin. The character of the country along this stretch remained more or less the same. Sand hills overgrown with pine and oak alternated with small fertile valleys. The prospects for a corn crop were excellent everywhere. Large farms tilled by negroes and small ones by free white labor followed each other in quick successions.

We spent the night at a farm several miles this side of Austin and did not enjoy the view of the capital of Texas until the following morning. A small, grassy, level plain, about one and one-half miles long and scarcely half as wide, bordered on the one side by a gently sloping chain of hills about one hundred fifty feet high from which several little brooks issue forth, and on the other by the forested bank of the Colorado, was chosen as a site for the city, and indeed a more beautiful and pleasant one could not have been selected. As to beauty of location I should prefer Austin to all other Texas inland cities, excepting New Braunfels and San Antonio. The houses lean so pleasantly against the slope of the hill on which isolated groups of live oaks are scattered about in the charming irregularity of natural primitiveness, that one would conjecture involuntarily that here indeed are the habitations of human happiness and peace.

This idyllic view from the distance is not wholly in keeping with the interior of the city. It was composed at that time of about one hundred to one hundred fifty frame houses painted white and a few log houses of which not a few showed unmistakable signs of neglect due to torn off boards and missing shingles. A single short but broad street was lined with houses, all other houses were scattered in irregular fashion in the prairie, although a very regular city plan had been outlined on the map.

The capitol in which Congress met formerly and where, since annexation to the United States, the Legislature of the State meets, is a log house on top of a hill, commanding the city. A more unpretentious building for a law-making body could hardly be found anywhere.

The Legislature had ended its annual session a short time prior to our arrival. This explained the deserted appearance of the place which reminded me very much of a small German bathing resort after the close of the season. For with the departure of the Legislature the rowdies, loafers, viz., adventurers, gamblers and hoodlums which drift in at the time of the sessions, had also left the city. The numerous grog shops were therefore deserted as they depended chiefly upon these people for patronage.

The city was founded in the year 1839, and was designated as the capital, which up to this time was located at Washington on the Brazos and prior to that time at Houston. It was thought that by selecting a place so far from the coast and on the extreme borders of the settlements and by founding a strong city, the other settlements lying downstream on the Colorado would be protected from the Indians, and that this city at the same time would form a base for further advancement of settlers to the north.

This purpose was indeed served to a great extent. The location of the capital in the wilderness, now the borders of the Indian territory, however, had many disadvantages for the legislators and officers particularly during the first years. It happened several times that members of Congress had their horses stolen at night in the vicinity of the city by the Indians and an open attack by them was not at all out of the question. Even today it is not advisable to venture more than two miles north of the city without being armed.

We found lodging in the home of a German who had lived in Texas many years and with whom my companion was formerly acquainted on his Aransas Bay farm. The memory of my stay in that house built of pine boards is, however, not one of my pleasant recollections, for we were troubled so much during the night with bedbugs, which had never before bothered me in Texas, that I, despairing of sleep, finally took my woolen blankets in order to lie down in the prairie, where our horses were grazing in enviable contentment. Here indeed I found a peaceful place the rest of the night under the beautiful moonlit sky.

On the next day we made a trip north of the city. About two miles distant from the city, a beautiful rounded hill, probably eight hundred feet high with sharp outlines and a heavy growth of cedar on its slope, presented an unusual sight. The Colorado issues from among these hills in a manner similar to that of the Guadalupe at New Braunfels. As a matter of fact, the location of New Braunfels and Austin are very similar. Both places lie on the border between the undulating hilly country and the hilly plateau, at the present time inhabited only by roving bands of Indians.

Later we also visited the camp of a company of dragoons which had been stationed near Austin since the annexation of Texas to the United States. The camp was situated in a pleasant location at a brook near the city. The officers and privates lived here in linen tents throughout the year seemingly without any other discomfort than the tiresome monotony of such a life. All troops of the United States are acquainted with it, for even during peacetime there is no entertaining garrison life for them in a resident city with its varied pleasures and sociableness. Their quarters are in most

cases a lonesome fort on the border where they are dependent for companionship on their own small circle.

My companion recognized among the mounted men and
horses several acquaintances from Florida, where this same
regiment had fought formerly in the war against the Seminole Indians.

In the afternoon I occupied myself with the geological examination of the neighborhood. The hill against which the
city is built is composed of white chalk marl, not very firm,
which is distinctly disclosed in the ravines made by several
small brooks. The fossils found here are of the same kind
found in the similar but more firm cretaceous formation of
the bed of the Guadalupe near New Braunfels.

On the following morning we started on our return journey to New Braunfels which was about fifty miles distant
from Austin. A young American lawyer from San Antonio
de Bexar joined us, for at that time the journey was considered somewhat dangerous as there were long, uninhabited
stretches along the route and wherever possible the journey
was made with companions.

After crossing the Colorado on a ferry and after the rather
narrow bottom of the river had been left behind us, our journey led us over treeless, grass covered hills, composed everywhere of chalk marl. Toward the south and southeast the
immeasurable, undulating prairie could be seen, whereas in
the north and northwest the wooded chain of hills arose.

At noon we rested at Live Oak Springs, a beautiful spring
surrounded by a group of live oaks and low shrubs. At this
same place Captain von Wrede and Lieut. Oscar Claren
from Braunschweig, formerly officers in Hanover, were
murdered by the Indians in October of the previous year,
1845. The particulars of this sad occurrence, which created
great sympathy throughout Texas, especially among the
German settlers, are as follows:

The two gentlemen just named, in company with a third
companion named Wessel, while on their return trip from
Austin to New Braunfels camped at the aforementioned

spring toward sundown and were just in the act of preparing their evening meal, when suddenly a band of thirty to forty Indians burst out of the bushes surrounding the spring and pounced upon them with their customary Indian war-whoop. The last named of the three travelers was some distance from the other two and after seeing his companions outnumbered, surrounded, and immediately fall though resisting bravely, he fled on foot to the prairie, since his horse had bolted upon seeing the Indians. He luckily escaped after out running one Indian and killing the other with a rifle shot. After walking about in the prairie for two days, he finally reached an isolated settlement on the Colorado in a pitiable, half-starved condition, from where he re turned to Austin. Accompanied by a company of mounted rangers he returned to the scene of the attack. Here they found the dead bodies of his companions, pierced with many arrows, scalped and stripped. They were buried at the place where they had fallen victims to the attack of the blood-thirsty savages.

The identity of the perpetrators of this murderous deed was not definitely established. It was generally ascribed to a band of Wacoe Indians who were also accused of horse theft at a number of isolated settlements. The fact that the Indians who attacked the men were not mounted would argue in favor of accepting this supposition, as the Wacoes, unlike the other tribes roaming in Texas, are never mounted.

The course of the San Marcos, the only important river between the Colorado and the Guadalupe was indicated from a distance by a narrow strip of forest a little beyond the other side of the Live Oak Springs. Soon after, we descended into the beautiful fertile valley. Before arriving at the San Marcos we had to cross the Rio Blanco, a pretty, clear stream, having its source higher up in the hills and uniting with the San Marcos a little farther down. A bottom, three miles wide, covered partly with brush, partly with luxuriant trees, separates it from the San Marcos. The latter is a beautiful river abounding in water, which flows rapidly and is of such magical clearness as can be found only in the rivers of western Texas. The springs of the San Marcos are only several hundred feet above the ford. Surrounded by the evergreen bushes of the palmetto (*Sabal minor Pers.*) and shaded by the stately forest trees, to whose pinnacle the mighty grapevines climb, resembling anchor ropes, they break forth under the thick limestone boulders with such tempestuousness and volume of water that they could turn mills at their immediate source. In this respect also the San Marcos resembles his neighbors, the beautiful Comal and the San Antonio rivers.

Unfortunately the broad fertile bottom of the San Marcos will never be suitable for agriculture since it is subject to inundation. While riding through it we noticed with astonishment dry cane and limbs hanging fifteen to twenty feet high above the ground from the trees, which the spring floods had left there.

The distance from here to New Braunfels is sixteen miles. The road led again over treeless, grass-covered hills. We, however, did not reach New Braunfels on the same day. We spent the night at York's Creek, an unimportant little brook. Here we met with an accident which happens quite frequently and belongs to the "petites miseres" of traveling in Texas. Upon awakening on the following morning my companions missed their horses. We looked for them in all directions and after becoming wet up to the waist while wading in the high, dew-laden grass, the tracks finally led us to the muddy crossing of a brook. We finally concluded that the missing animals had departed for New Braunfels without their masters.

We had no alternative but to load all our baggage on my mule which I had prevented from running away by tethering it during the night, and to trudge nine miles to New Braunfels. In this pathetic manner, I leading the mule and my companions taking up the rear, we finally reached our destination. The two lost horses, moreover, were recovered two weeks later in San Antonio by their owners, just as they were being led away by a company of volunteers going to the Rio Grande. They had been caught by Mexicans and fraudulently sold to the volunteers.

Chapter XIV

Weather Conditions—Volunteers from Lamar County—
Election of Comal County Officers

T he weather was beautiful throughout the entire trip. During the hours from twelve to four o'clock the temperature registered 75 °F., and in the morning and evening hours 66° to 72 °F. This temperature was maintained until the middle of June. During the second half of this month we had thunderstorms almost daily. Before discharging, they would usually threaten for a long time and then the air was sultry and oppressive. The discharge followed with great intensity. The reverberating of the thunder was grand. The torrential rains, accompanied frequently with hail, came down with such force and volume that one could not understand one's own words due to the noise occasioned by their falling upon the wooden roof of the Verein's building. The plain upon which the city was built was often converted into an inland sea within a few minutes. The air became somewhat cooler after these thunderstorms, but the temperature never dropped blow 65°.

The Guadalupe would often rise fifteen feet above its normal stand after these heavy rains, carrying with it in its swift torrent a number of large trees, uprooted farther up in the hills. Smaller brooks, ordinarily not containing flowing water, became raging torrents which could be crossed only by swimming.

The Texans considered this weather very unusual, since it is usually dry at this season of the year. This weather made communication with the interior very difficult, particularly for transporting the German immigrants from the coast to New Braunfels. All the roads in the level lowlands along the coast become bottomless and farther inland a single creek often detained wagons and people for weeks.

On June 28 several companies of mounted volunteers, four hundred men strong from Lamar County on the upper Red River, passed through New Braunfels on their way to

Mexico. They were wild, desperate looking, rough fellows, clad mostly in deerskin. Similar to all Texas volunteers, they carried the long American rifle, and in addition revolvers and bowie knives. Since the ferry had been swept away and only a small boat was available, they were obliged to swim their horses across. Sitting entirely naked on them, whole sections plunged with loud "hurrah" into the swift and still swollen stream. After thus swimming beyond the middle of the stream, where they could be assured that the horses would continue to the opposite shore, they slid off their mounts and swam back in order to get the rest of their equipment. All this presented an animated and attractive picture, enhanced by the natural charm of the surroundings at the juncture of the Comal and Guadalupe.

The 4th of July, the great national festival celebrated in commemoration of the signing of the Declaration of Independence in the year 1776, was also celebrated here. A large American flag was hoisted on the Verein's building and a formal banquet was given to which the officers of the Verein and a number of dignitaries of the city were invited.

The heat had increased considerably during the first days of July and at noon the thermometer registered between 79°F. and 86° F. However, I must confess that it never became oppressive and disagreeable to me. Of course, I refrained from leaving the house during the hottest hours of the day. I was also fortunate in that the house was on a hill, where throughout the day the south wind blew strong enough continually to carry off paper and other light articles through the open door.

As a rule, I spent my noon hours in reading, sitting in the dining room between the open folding doors facing north and south. At no time did the heat hinder me from sleeping at night, whereas in New Orleans in the following year in April the nights appeared unbearably hot to me.

Unfortunately the state of health of the inhabitants grew worse with the increasing heat. Dysentery, in particular, spread in such a virulent form, that two and three deaths occurred daily in New Braunfels. This sickness was especially prevalent among the newly arrived immigrants from Indian Point, who had contracted the disease during their stay on the low coastal country or on the long tedious journey, but the older settlers also were not spared.

On July 13 the inhabitants of the city became greatly excited. The first election of county officers took place. The territory in which New Braunfels lay belonged to Bexar County up to this time, but by an act of the last Legislature it had been created into a new county called Comal.

The following officers were to be elected: a chief justice, a district clerk, a sheriff, a justice of the peace, a coroner and a probate judge. All were permitted to vote who were citizens of Texas and more than twenty-one years old at the time when the United States ratified the Constitution of Texas and who were residents of the county for the past six months. The choice for the two most important offices, that of chief justice and district clerk, fell upon two Americans, or more correctly stated, Americanized Pennsylvania Germans. It was of course almost imperative to select them, since among the immigrated Germans very few had the necessary knowledge of the English language and even less of the laws and order of business.

On July 15, Herr von Meusebach returned to New Braunfels with several companions from Nassau after having been absent almost eight months. This was a great event for the whole city because his return gave rise to the hope that the financial situation of the Verein had now improved. At intervals during his absence, comparative small amounts of money had arrived at New Braunfels. These amounts afforded only temporary relief, owing to the heavy expenditures of the Verein's management, which consisted in advancing money for the support of a great number of immigrants in New Braunfels as well as in Fredericksburg, further, for transporting the people from the coast to New Braunfels, for salaries of the officers, etc. In the meantime bills of exchange and promissory notes took the place of money. But when promissory notes were not paid when due and the bills of exchange drawn on the mercantile houses in New Orleans and Galveston were refused, difficulty was also experienced with these makeshifts. The creditors of the Verein refused to accept such bills of exchange and promissory notes for their claims, therefore they could not be used at all in ordinary business or they were greatly discounted.

For a long time money was scarcely to be found in New Braunfels and most of the mercantile and other establishments transacted business by crediting or debiting the accounts of the contracting parties in the ledger of the Verein's treasury.

Unfortunately this condition was not entirely remedied upon the arrival of Herr von Meusebach. However, he did bring along several thousand dollars in cash, which raised the hopes of all persons interested in the continuation of the Verein, since he was able to satisfy many small creditors and make partial payment to the larger ones.

At the same time energetic arrangements were made to house and care for the sick among the immigrants who had arrived from the coast and who up to this time were obliged to live in the open. A long shed was built on a suitable place near the Comal in which the sick were housed and daily visited by a physician. Another physician, Dr. Remer from Breslau, was engaged to bring succor to the sick immigrants on their way from Indian Point to New Braunfels. An orphanage was also established in which the many poor children, who had lost their parents in the epidemics, were taken care of.

In fact, everything was done to help the prostrated cause of the Verein as much as possible and to ameliorate the suffering caused by the delay in transporting the immigrants from the coast to the interior. A complete cure of the mistakes from which the Verein enterprise suffered was out of the question as far as the persons who had charge of its management in Texas were concerned.

CHAPTER XV

On July 24, I again took a trip of several weeks' duration, into the interior. The trading post of the Torrey brothers (Torrey's Trading House) on the upper reaches of the Brazos was my destination. This is the only place authorized by the government where trading with the different Indian tribes of Texas may be carried on. My companion was no other than one of the owners of the trading post, Mr. John Torrey of Houston. The advantages and amenities promised by the companionship of this man, personally acquainted as he was with all the conditions of the country and with the peculiarities of the various Indian tribes through frequent contact with them, counterbalanced all scruples which arose in me when I considered that we were at that time in the hottest period of the year in which the thermometer registered between 90°F. to 100°F. during the day, and that travel for the Europeans not fully acclimated was always precarious.

Our first halt for the night was made at the San Marcos, since we had left New Braunfels in the afternoon. Everything appeared more animated at this place than on a recent visit, for a company of mounted rangers was stationed here, and several families with large wagons had arrived to found a new settlement. In fact, a more advantageous and pleasant place for a settlement could not be imagined than this park-like little prairie, surrounded on one side by the forest fringing the San Marcos and on the other by the steep hills, the beginning of the higher hill country.

We chose our night's lodging with the rangers since it saved us the trouble of kindling a fire and preparing supper. The camp was close to the banks of the Comal among high trees, under which the underbrush had been cleared. A number of linen tents had been erected in irregular order, which would probably have afforded the inmates only partial shelter dur-

ing a heavy rain. But upon examining the inmates a little
more closely, one could readily see that they regarded such
inconveniences as mere trifles. All were wild, rough fellows,
sons of settlers, who had exchanged the work in the field for
a time for the irregular life of a soldier. Their mode of living
was simple enough. The victuals furnished them now were
limited to dried beef, cornbread and coffee. However, they
expected to be supplied soon with better provisions and to
be reimbursed with money for the previous deficiencies.

When I awoke on the following morning, I made the
unpleasant discovery that the same misfortune which had
befallen my two companions on a former trip to Austin, had
also befallen me. My mule, which I had tethered to a tree
near my sleeping quarters, becoming dissatisfied with the
sparse grass which the horses of the rangers had cropped
closely, had gnawed the rope in two. The tracks indicated
clearly that it had decided to return to New Braunfels. Fortunately, by promising him three dollars, I induced one of
the rangers to pursue and capture the fugitive. Toward noon
he returned with the mule which he had found quietly grazing six miles this side of the Guadalupe. Thus this misfortune caused only the loss of several hours.

Since my companion had to go to Bastrop on business,
we left the road to Austin after crossing the Rio Blanco and
turned somewhat east. The open prairie began again after
crossing this stream. Our road first led us over a prairie
several miles wide, the so-called Hog-Wallow Prairie, viz.,
a prairie whose surface is covered with countless flat, irregular depressions about five to six feet in diameter. Many
of these are found in all parts of Central Texas. The cause
to which the depressions owe their existence has so far not
been definitely determined, although several solutions have
been offered.

Our camp on the next night was made about fifteen miles
on the other side of the San Marcos, at a waterhole called
by the uninviting name of "Alligator Hole." Alligators, however, did not molest us, but swarms of mosquitoes fell over

us as soon as we lay down. My apprehension that Indians would be attracted by our fire did not frighten my companion in the least. He traveled entirely unarmed, depending for protection upon his personal acquaintance with most of the Indians traveling about on the borders of the settlements. With reference to our rather simple supper, he exclaimed almost resentfully, "... — ... I wish there were Indians here, we might get some venison from them!" A pack of prairie wolves, not at all dangerous, kept up a continuous howling near us throughout the night.

Just as we departed the next morning at sunrise, we frightened away a herd of mustangs or wild horses which had come to drink in a neighboring water puddle. A magnificent white stallion led the herd as they sped away.

The prairie, which extends from the San Marcos, hardly interrupted by the very small narrow strips of forest along several brooks, ends about ten miles this side of the Colorado, and sparse forests alternate with small plains overgrown with scattered mesquite trees. At the same time farms also appeared again of which we had not seen any since leaving the San Marcos. A broad strip of sterile soil composed of gravel which is covered with a monotonous post oak forest, is found near the Colorado.

The Colorado at Bastrop is enclosed by a very small bottom. In it, the trumpet vine, which is found in all river bottoms of Texas and which is also cultivated at home as an ornamental plant, had just opened its scarlet flowers.

We spent the night in Bastrop and continued our journey the following morning. Our next destination was the hamlet Caldwell on the Brazos, about sixty miles distant. We followed the old Spanish military trail, the Presidio Road, which leads from Presidio de Rio Grande on the Rio Grande across Texas to Nacogdoches. When speaking of this road, one must not imagine a highway in the European sense; it is primarily a road to indicate directions, sometimes very indistinct, dim wagon trail or horses hoof prints in the high grass only showing the direction it takes. In avoiding all natural obstacles and pointing out the best fords over the rivers and creeks, it is of great value to the traveler.

The road at first led us over sand hills one hundred to one hundred fifty feet high. They resembled those I had learned to know along the entire way from La Grange to Bastrop. A beautiful pine forest covered the sand hills. Such pine forests are not found again west of here as far as the Rio Grande. On the other side of the hill where the undulating ground is less fertile, post oaks take the place of the pines, and for the next forty miles form at first a continuous forest which later on alternates with small prairies. We rode all day through this monotonous forest without seeing a house, and made our camp in the evening on the banks of the Yegua, a sluggish, muddy river enclosed by a broad forest bottom. I shall never forget this camp on the Yegua, for nowhere was I molested more by mosquitoes than there. No sooner had I lain down under a live oak tree, tired and worn from the long trip, when I heard their ominous humming. I tried to protect myself against their stings by pulling my woolen blanket over my head, but alas, it was impossible to remain under this cover any length of time, for the night was warm. The hated insects also managed to find their way under the cover. They stung me mercilessly. After spending sleepless hours, I finally sprang up in desperation, resolved to spend the rest of the night by the fire. Upon nearing it, I saw my companion sitting there. Without questioning him I knew from his imprecations that he had not fared any better than I had. We decided to leave this confounded place as quickly as possible and by the light of the stars we saddled our horses to resume our journey.

The forest ended about fifteen miles this side of Caldwell and we entered a wide prairie, called the San Antonio Prairie. On its rim, we saw isolated farms—the first we had seen since leaving Bastrop.

Caldwell is a small place containing about thirty to forty houses scattered about. It has the reputation of being an unhealthful place and is without any particular attractions. Just as we arrived there, we saw a number of fashionably dressed women coming out of a log house, standing on an elevation. They mounted their horses tied to nearby trees and departed in various directions in company with a few men.

We learned from an acquaintance whom my companion met, that this was a Baptist meeting which had lasted three days. In the sparsely settled regions of Texas and the western part of the United States, these religious meetings have almost the importance of a holiday, of which there are none except the Fourth of July. Everybody is glad to go to such meetings since it gives distant friends and acquaintances an opportunity to meet.

We remained long enough in Caldwell to avoid traveling during the hot noon hours, and then rode several miles to the home of a wealthy farmer, a Mr. Porter, from whom my companion, Torrey, wanted to buy beef cattle. This man had five hundred head and offered them for sale at two cents a pound with the stipulation that the buyer transport them himself. This added condition was not of little importance since a good many of these half-wild animals, raised miles from human habitations, escape and are lost when driven to places several days distant.

After supper, preceded by a long, spoken prayer and the reading of a chapter of the Bible in the presence of the entire household, the slaves included, Mr. Porter related many a tale about the trouble he experienced with the Indians in earlier times as he was one of the first settlers. One of his relatives was murdered by the Indians near the house and he himself had not gone to the field for ten years without his gun.

A ride of eight miles through an open prairie, with only here and there oak groves, brought us to the Brazos on the following morning. It is the same muddy yellow stream here as found at San Felipe de Austin and confined to a narrow bed by high banks. After crossing the Brazos over a ferry and a few miles farther the Little Brazos, a shallow but rather broad creek with a rocky bed, we soon came again to an open prairie alternating with oak groves.

We spent the night with a wealthy planter who owned a considerable number of slaves. Despite his wealth, the house of this man was a simple, unornamented log house, similar in style to the farmhouses of Texas. The three daily meals consisted also of the same kind of food previously mentioned and there was no change in the menu. It is remarkable that the American, though otherwise in agreement with his English cousin, does not share with him the love for a comfortable and cozy home.

With reference to the young and rather pretty wife of the planter, my companion informed me that she had many suitors before her marriage, for in addition to outward attractiveness, she was also "ten or twelve negroes' worth." Thus do the negroes form the basis for gauging the wealth of a person in this country.

The man informed us that it was his intention to abandon this farm, although beautifully situated, and to buy farther down on the Brazos because it cost him too much (about $500.00 annually) to move his cotton to the Houston market. The scarcity of easy communication and especially natural waterways, of which other states of the Union have an abundance, is so general in most parts of Texas, that it proves an obstacle in the development of the country.

On the following morning we rode toward Wheelock's Settlement about thirty-five miles from Caldwell and two miles from the Presidio Road. The settlement consisted of about a dozen houses located on an elevation surrounding this section and of all those I saw in the upper Brazos region, this settlement pleased me most. We paused at the home of a merchant, Mr. Kellock. This man had a well-kept garden near his house in which a variety of garden plants grew prolifically, an unusual sight in Texas. Here for the first time I also saw okra grow, whose half-ripe fruit capsules furnish a very tender, palatable vegetable. There was also a beehive of peculiar construction in the garden, of which kind I later saw some more in Texas. This was a small enclosed house built of boards, about ten feet long and six feet wide with several horizontal partitions within. Through a number of small holes, the bees have access to the inner rooms and many swarms are busy at the same time in the same house with no other separation but the horizontal boards placed therein. Such a so-called bee palace is superior to the commonly used individual bee hive since the honey and wax can be removed with greater ease.

Before we continued from here to resume our journey toward the north we made an excursion to Boonville, which lay about twenty miles south from Wheelock's Settlement. My traveling companion had business there and I accompanied him, since we heard that, in a creek near the place, whole trunks of petrified wood were to be found. The way led through an almost level, open prairie which was broken only occasionally by oak groves. The place itself, county seat of Brazos County, consists of about a dozen houses lying in a small oak grove and offers nothing unusual; nevertheless of interest to me was the presence of petrified wood in the vicinity. The place where it is to be found is about two miles from this spot. A small brook had cut several feet into the diluvian gravel and sand and in the bed a large number of pieces of petrified wood of various sizes appeared whose former texture was very plainly preserved. I selected a single block, four feet long, representing the lower part of a trunk about three feet thick, because of its good preservation, with the intention of taking it to Houston from which place I later took it with me to Europe. The settlers in Texas erroneously take this fossil wood to be petrified post oak and believe that it petrified at the place where it is found. But it belongs rather to a species of wood which no longer exists in Texas and usually shows traces, like the rubble of the gravel in which it is found, that it has been carried from a more distant point. The prevalent legend, also repeated by Kennedy of a petrified forest in upper Texas, no doubt is also based on a similar occurrence of an accumulation of petrified pieces of wood.

In the immediate vicinity of this place where these large blocks of petrified wood occur, there is found in the bed of the creek, covered several feet with gravel, the same tertiary formation which I had noticed previously at the crossing on the Brazos.

On our return trip to Wheelock's Settlement, we stopped at the house of a farmer, who had such an abundance of ripe peaches that he gave bushels of them to his neighbors who came with pack horses to carry them off. He regretted that since the trees stood in his cornfield, he could not drive his hogs in to feed on them. We observed a similar abundance of this fruit at several other places, and found some of excellent quality.

On the following day we continued our journey from Wheelock's Settlement to the trading post. The region from Wheelock's to the hamlet, Bucksnort, about forty miles distant, which lies opposite the falls of the Brazos, is monotonous and not very attractive. Post oak forests and occasional little prairies predominate. The soil is mostly sandy and the sluggish, muddy streams, usually entirely dry in summer, irrigate the land but poorly.

With the exception of the hamlet Franklin, ten miles from Wheelock's Settlement, containing about a dozen rather dilapidated log houses, we saw at long intervals, miles apart, a few farm houses, usually built on the rim of the creek bottoms, whose black humus had offered an inducement for a settlement.

That night we lodged with a farmer in Bucksnort whom my companion had known in former days. He told us many stories of the trouble he had to suffer in former years on account of the Indians, when his home was the farthest advanced settlement on the upper reaches of the Brazos. At one time he was actually besieged on his farm and for such an event the various log houses were arranged in a manner, that the windows and doors opened into an inner court thus permitting them to defend one another.

Here, by the way, we partook of the simplest supper which I had eaten so far on my entire trips in the country. It consisted of nothing else but cornmeal mush and cold milk.

Our host offered his excuses with a certain amount of embarrassment for his inability to entertain us properly. Later my companion informed me that this man was well-to-do at one time and that he always had a surplus of food. Because of a passion for racing, for which he, himself, kept expensive horses, and on which he had placed high bets, he was ruined financially. The same thing had happened to other inhabitants of the place. To hear people speak of racing on the extreme outposts of civilization, sounded peculiar to me, but on the following day I actually saw a racetrack which had evidently been recently used. Later I had other occasions to observe, that the love for this national sport asserted itself in these places, far removed from civilization.

Our lodging place for the night was just as unusual as our supper. When we expressed the desire to lie down, the sons of our host explained to us that it would probably be better to follow their example and sleep on the roof of one of the outhouses since it was probable that mosquitoes would molest us too much near the ground. We were not inclined to doubt the latter supposition as we heard the ominous humming of these little pests during the course of our meal. Following their advice, we took our woolen blankets and climbed on the gently sloping roof of the house pointed out to us and, although our bed was rather hard, we actually slept there unmolested by the mosquitoes which remained closer to the ground.

In the afternoon of the following day we left this farthest settlement on the Brazos (which evidently was not in a prosperous and growing condition) in order to reach Torrey's Trading Post, about twenty miles distant. Just north of this place the character of this country changes. The ground rose to higher elevations and instead of dense forests, we saw extensive prairies covered with mesquite trees, but now and then we also passed oak groves. In general the topography of the country assumes the appearance of the region between Austin and New Braunfels. Toward the west one can see the level wooded bottom of the Brazos extending for miles.

We, however, did not quite reach our destination on this day, but camped in the open prairie several miles this side, in order not to lose our way in the darkness. Our horses found excellent food under the mesquite trees, but we, ourselves, were entertained before going to sleep by the beautiful spectacle of a prairie fire. Like a sparkling diamond necklace, the strip of flame, a mile long, raced along over hill and dale, now moving slowly, now faster, now flickering brightly, now growing dim. We could the more enjoy this spectacle undisturbed, since the direction of the wind kept it from approaching us. My companion was of the opinion that Indians had without doubt started the fire, since they do this often to drive the game in a certain direction, and also to expedite the growth of the grass by burning off the dry grass.

On the following morning after a short ride of a few miles, on rounding a corner of the forest, we suddenly saw the trading post before us. It lay on a hill covered with oak trees, two miles distant from the Brazos, above the broad forested bottom of Tohawacony Creek. The entire layout consisted of from six to seven log houses built in the simple customary style out of rough, unhewn logs. These houses were not surrounded by palisades, as are those of the trading companies on the Missouri, neither do they contain any other protective enclosures. The safety of this trading post against possible Indian attacks is founded rather on its usefulness, in fact its necessity to the Indians.

The largest of the log houses contained the pelts received in trade from the Indians. Buffalo robes or buffalo rugs and the hides of the common American deer (*Cervus Virginianus* L.) formed by far the greatest number of hides. Some of the buffalo skins are brought to the trading post entirely raw, some are tanned inside only, and very often they are painted more or less artistically. Their value depends upon the size, the uniformity of the fur, and also upon the artistic paintings on the inside. The hides of average quality were sold in Houston for three dollars, and the fancy ones for from eight to ten dollars. Most of them are shipped to the Northern

States, and also to Canada, where they are used for covers in sleds or for wagon seats. Leather is not made out of the skins since it is too porous and not compact enough.

The number of buffalo skins brought annually for trade from the western prairies of North America is estimated to be over 100,000. The number of animals killed for sport or just for their meat is no doubt much larger for hardly half of the year is suited for the treating of skins. Diminished annually in such considerable numbers brings home the thought that their extermination is not far distant. The huge expanse over which they are scattered is no argument, as Gregg correctly observes in *Commerce*, against the accepted fact of their imminent total extinction, especially when one takes into consideration that they have already disappeared entirely from a large portion of the continent which they formerly inhabited. Old settlers, still living, recall that the buffalo was almost equally plentiful east of the Mississippi in Kentucky, Indiana, etc., as he is now on the prairies between the western boundary of Missouri and Arkansas and the mountain range, and according to historical data it is certain that he once inhabited the Atlantic coast.

Because of their soft pelt and the most artistic painting on the inside, I selected a number of skins of buffalo calves. The value of all other skins brought by Indians for trade is comparatively small; they are skins of the raccoon, the cougar (*Felis concolor L.*), also occasionally of the beaver, the antelope (*Antilope Americana Godman*), bobcats and gray wolf (*Lupus occidentalis Rich.*). Aside from the skins, mules are especially an important article of trade with the Indians. Most of them are captured by the Comanches on their annual raids in the northern province of Mexico and the larger number of these are wholly wild when brought by the Indians for trade. During our stay, the taming of about fifty of such mules which the Comanches had traded in just shortly before, was in progress. On this occasion we saw with what dexterity two Mexican lads, who had lived a long time among the Comanches as captives, lassoed the mules in the corral, excelling even the adult Americans who also tried. Among the Mexicans the practice in the use of the lariat, which in a great measure serves both as tool and weapon, begins in the earliest youth and this accounts for the unbelievable dexterity that they display. G. W. Kendall relates in his interesting report on the ill-fated Texas expedition to Santa Fe, New Mexico, that he frequently saw little Mexican boys lassoing the chickens. They were able even to lasso the particular leg they desired.

When tamed, the mules are worth about $40.00 on an average. At the time of my stay a number were sold to a company of rangers at this price who used them as pack animals.

Another log house contained the goods which the Indians receive in exchange for the articles they bring. The most important of these are: woolen blankets, coarse woolen cloth (called strouding) colored especially scarlet and blue and used in manufacturing breechclouts, printed calico for shirts, also thick copper wire to serve as ornaments for arms and legs, knives, glass beads, powder, lead, tobacco, etc.

The remaining log houses contained the dwellings for the various persons who resided at the trading post. At the present time these consisted of an agent in the employ of the Torrey brothers who knew how to trade with Indians; furthermore a gunsmith appointed by the government who repaired the guns of the Indians with which at least a number were provided; and finally an old trapper or fur hunter who on account of gout and rheumatism had become unfit for that sort of life. He had recently made this his home in order to be as near as possible to the wilderness, the scene of his former joys and deeds, and as far as possible away from the hated civilization. This old trapper was ever eager to relate to anyone who would listen, the tales of his lonesome beaver hunts in the mountain region and other incomparable charms of a trapper's life.

In addition there was also an Indian whose duty it was to beat the pelts in order to rid them of insects. To watch this one at work always furnished me entertaining amusement, for his face told with each lick how the performing of such menial work for the pale faces was repulsive to his national inclination for laziness and his consciousness of Indian dignity.

Finally there were two Mexican boys, seven and nine years old, previously mentioned. These had been captured about a year and a half ago, from the Mexican settlement on the Rio Grande and had since then lived among the Comanches until recently when the boys were brought by them to the trading post for sale. The kind-hearted owners of the post had ransomed them.

Since parents or other relatives could not be located, they remained here for the present. However, just at the time of my stay, another Mexican youth who had reached the trading post under similar circumstances, was reunited with his overjoyed brothers who had made the journey from Presidio de Rio Grande for that purpose. The $120.00 ransom paid for the boy's release from the Comanches was returned.

In addition to the persons mentioned there was also another agent in the employ of the government (a so-called Indian Agent) who maintained regular quarters at the trading post but who at the time of my visit happened to be on a trip. This agent was well versed in the languages and customs of the various tribes whom he visited from time to time, serving as interpreter in concluding contracts and peace treaties. He also informs the government of their wishes and needs.

The mode of living on this outer fringe of civilization is in harmony with the wilderness and primitiveness of the surroundings. Dried buffalo meat, smoked buffalo tongue, considered a delicacy in the civilized parts of the United States, bacon, honey and bread were considered the choicest food. A pile of buffalo hides formed an excellent bed.

Already on the second day of our arrival a small Indian band chanced to arrive in order to trade. To see the long train, resembling a caravan, appear on the hill on which the trading post stood, presented a picturesque and fascinating sight for a European. According to Indian custom, they rode single file, the men in advance, dressed in their best, looking about, dignified and grave; the lively squaws following, sitting astride like the men, each usually carrying a black-eyed little papoose on her back and another in front of the saddle. At the same time they kept a watchful eye on the pack horses which carried the skins and the various household goods. A halt was made near the houses and immediately the squaws were busy chopping down branches, needed for the construction of their tents.

Later the skins to be traded were brought to the warehouse, weighed, and the value determined. The Indians then selected goods equivalent to the price agreed upon. Usually such a visit lasted several days and had the same happy significance for the Indians as a country fair has for our German people.

Chapter XVI

VISITING THE CADDOE INDIANS—LARGE BUFFALO HERD—
GEOLOGICAL OBSERVATIONS—HIGH FEVER—BUCKSNORT

After spending a few days at the trading post, a longed for opportunity presented itself to become acquainted with the regions lying farther up the Brazos. My companion Torrey had to transact some business in a little place called Dallas on the Trinity River. Since I knew that in this direction there was little of interest to me, I preferred to accept the offer made by the previously mentioned gunsmith at the trading post, Mr. Cockswell, to accompany him during the absence of Mr. Torrey to an Indian village of the Caddoes about sixty miles up on the Brazos.

Our preparations were of the usual kind necessary for a several days' journey into the wilderness. We put a sack of coffee, salt and zwieback, made of wheat flour, into the saddlebag. A tin vessel was tied in the back of the saddle and besides this only a rifle was taken along, in the use of which my companion was master, as are most American settlers. The circumstance that Mr. Cockswell was personally acquainted with most of the Indians roaming about on the Brazos was sufficient guarantee in the event of meeting hostile Indians.

Our journey first led us through the bottom of the Tohawacony Creek, on the rim of which Torrey's Trading Post stood. It was several miles wide here although the creek is only a small stream in which nothing but a few mud holes are found during this time of the year. Several Indian families, the remainder of a once numerous tribe, lived miserably in huts made of branches in a very low place where the Brazos overflows annually. The high, dense trees do not permit the sun's rays to penetrate to the ground, thus creating an unhealthy atmosphere in which no European could live without becoming a victim of sickness.

After leaving this wooded bottom behind us, we entered a sparse oak forest which led us to an open, undulating prairie extending toward the north and east in an immeasurable distance. Just before entering this prairie, we met a young Indian who had just killed a buffalo cow and who was engaged in loading some of the choicest pieces of meat on his horse. He gladly permitted us to select several choice pieces of meat from his surplus in exchange for a handful of salt. Under the trees nearby we saw a buffalo calf which had lost its mother in the cow just killed by the Indian, running around crying pitifully. The poor animal showed so little fear of us in its grief that we could have easily caught it.

Having thus secured something for our supper without any effort, we selected our camp for the night near a small brook which was dried up with the exception of a few water-holes. In a short time we had roasted several pieces of meat on smooth staves placed slanting against the fire. In spite of its simple preparation the meat had an excellent taste. Now I could understand why the hunters on the western prairies preferred buffalo meat to all other meat. Indeed it can be compared to the best beef and has this additional advantage that the fat between the muscles is distributed more evenly. Of course it is not equally good at all times of the year and the age of the animal also makes a considerable difference. The flesh of an old, lean bull is almost unfit for consumption. The best meat is obtained from a fat, young cow and particularly the so-called hump ribs, streaked with fat, which form the base of the hump, is highly prized. The tongue, the udder, and the marrow of the long vertebrae roasted over the fire, are also considered delicacies.

When on the following morning at daybreak we entered the prairie on which mesquite trees grew scatteringly, the first object that met our view was a buffalo herd, quietly grazing near us. I had long cherished the wish to see this largest mammal of the American continent in its wild state. This desire had increased since I had seen so many tracks crossing our road the day before. Now all of a sudden a whole herd stood before us. Very likely not expecting a visitor at such an early morning hour, or rather not having been warned by their sense of smell (on which they depend more than sight) due to the direction of the wind, they permitted us to examine their ungainly bodies and clumsy movements in a leisurely way.

The American buffalo or bison (*Bos Americanus Guel.*) called "cibolo" by the Mexicans has, as is known, nothing in common with the buffalo of southern Europe, introduced from India, except the name. It is rather closer kin to the aurochs now found in a wild state in the woods of Bialowieza in Lithuania, which formerly roamed over all of Germany and the greater part of central Europe. It distinguishes itself from the latter on account of its smaller size, the comparatively weaker hind part of the body, and the long, shaggy hair on the head and neck. These characteristics give the animal a formidable appearance, which, however, is not in keeping with its timid and peaceful disposition, causing it never to attack but to seek safety in flight at the approach of an enemy. A peculiar osteological mark of distinction between the American buffalo or bison and the aurochs of Europe is the number of pairs of ribs. The former has fifteen, the latter fourteen.

After we had observed these huge forms sufficiently, we rode toward them and immediately an old bull stopped grazing, raised his matted head and bolted. The whole herd followed him in their peculiar, clumsy, but nevertheless fast gallop and in a short time all had vanished behind a swell in the prairie. The whole prairie was covered with countless buffalo trails, crossing in all directions, reminding one of a European grazing ground. During the course of the day, we saw more herds, numbering three to four hundred, some of which we were able to approach so near, that had it been our desire, we could easily have shot several. On another occasion, however, we saw them pass by on the run, probably because they had been disturbed by the Indians.

Our road, only an indistinct Indian trail, led us over open, level prairies throughout the day. These prairies were covered with excellent tender grass and here and there with scattered mesquite trees. Several clear brooks irrigated the land. The fertility of the soil and the high elevation which appears to exclude certain causes of local diseases, will no doubt invite the settlers, and human habitations will soon be found here.

The subsoil throughout this region is a white chalkmarl of like quality as that found in Austin, New Braunfels and especially prevalent in western Texas. At the crossings over several small brooks, this rock formation was well-exposed and I observed in it numerous fossils, which do not leave any doubt as to its age. Above all others the widely distributed *Exogyra arietina* was common. Just as at New Braunfels, I also collected *Gryphaea Pitcheri Morton* at the same time. In many specimen was also found a new, large species of the genus *Turrilites*, not observed at any other point.

After a ride of about thirty miles we saw, toward sundown, from the top of a hill, the destination of our excursion—the Caddoe Indian village lying before us. A more suitable and pleasant place could not have been selected by the red sons of the wilderness. The village lies in the center of a plain two miles long which on the one side is bordered by the wooded banks of the Brazos and on the other by steep precipitous hills. A beautiful clear creek flows diagonally through this plain on a smooth bed of limestone and along its banks are several large live oaks. The huts of the Indians stood on both sides of this creek in picturesque disorder and near each was a cornfield. Between the hills from which we looked down and the village proper, about a thousand head of horses were grazing in the plain, among which a number of naked, long-haired Indian boys rode back and forth yelling. Thereupon we descended to the village. At the various huts which we passed we were welcomed in a friendly manner by the inmates, as my companion was well known to all of them. We, however, refused the repeated invitation to sleep in one of their huts, for eager though I was to learn something firsthand about the Indian home life, we dreaded that close contact with the pesky little insect world, which inhabits every Indian dwelling. We preferred spreading our blankets under a live oak on the banks of the creek. Before we lay down to sleep, we were visited by several Indian squaws who brought us watermelons as a present, and who received glass beads from us in return.

Just after sunrise on the following morning we took a walk through the village. The home of every family consisted of several huts of diverse form. There is always a large conical shaped hut present, about fifteen feet high, which is enclosed on all sides except for a small opening at the bottom. It is thatched with long grass and therefore at a distance resembles a haystack of medium size. This hut is used in cold and wet weather. Near it are several other huts, open on the sides, which really are only grass-covered sheds, resting on four uprights under which at a distance of about two feet from the ground, a horizontal latticework platform, woven from strong twigs, is fastened. On this wicker platform men and women squat during the hot hours of the day. The roof shelters them from the hot sun's rays and at the same time the air can circulate freely on all sides, even from the bottom. Finally there was a third kind of hut used for storing provisions which was nothing but an oven-like container also covered with grass, resting on four high posts.

Adjoining each hut was a patch of corn and watermelons as was previously mentioned. The patches looked disorderly and uncultivated enough and no fence prevented the horses or other animals from entering them. The corn which they cultivate was a quick maturing variety with small kernels.

The well-informed Gregg divides the Indians of the western prairies of North America into (1) civilized Indians, such as the Cherokee, Creeks, Choctaws, etc., which live in regular villages, carry on agriculture and other crafts, and possess a sign language; (2) intermediate tribes, who combine agriculture with the hunt, and, at least part of the year, live in more permanent houses; (3) finally, wild tribes, such as Comanches, Apaches, Arrapahoes, etc., who live by plundering, and who, without a permanent place, erect their wigwams here and there.

The Caddoes, small in number, belong to the intermediate tribes, as do also those living farther north, such as the Pawnees, Chippewas, Potawatomies and many others. They have their residence chiefly in northern Texas and live together with two other small tribes, the Keechyes (Kihtscheis) and Inyes (Eineis). This was also the case at the village visited by us. According to Gregg, the Caddoes and Inyes number only about five hundred souls.

Despite the early morning hour we found already, in the first hut which we visited, all the inmates, including even the smallest urchin, busily engaged in consuming watermelons (*Cucumis citrullus L.*) which apparently had not been given time to ripen fully. We found the same condition in all other huts. It seems indeed as if watermelons comprised almost the sole food of the Indians at this time of the year and they were consumed in incredible numbers; a mode of living which would undoubtedly prove fatal to the whites living in the country. We found the Indians in the best of humor everywhere, and my companion assured me that these sons of nature live in harmony and that bickering and strife are unknown.

At all the huts we visited, everybody wanted to know how many buffaloes we had seen and where they were to be found. As a result of our information several companies left during the course of the day to hunt them. Women and children and a number of pack horses to carry away the meat as well as the pelts were taken along in joyful anticipation of a successful hunt. Our appearance in the village did not create much excitement which indicated that these Indians had frequent intercourse with the people of the settlements. Their dress also spoke of their contact with the civilized world, since the various component parts were evidently of European origin, particularly the cotton shirts with which most of them were supplied.

It was our intention to extend our excursion farther north where the land appeared to be of a different character and much wilder. Higher, sharply outlined hill formations arose and the Brazos, similar to the Guadalupe above New Braunfels, flows through a narrow gorge whose perpendicular, dazzling white limestone walls were plainly visible from the river. However, in the course of the day I became indisposed. In a short time this sickness developed into such a virulent climatic disease that my life was in danger and its effects did not leave me until my departure from Texas in the following spring.

Up to this time I had felt exceptionally well without even experiencing those little discomforts, such as swelling of the feet and eruptions, which usually accompany acclimatization of the European in Texas. The extreme heat of the day, registering 90°F. to 100°F. to which we were directly exposed continually during the previous days was no doubt the immediate cause of my indisposition. It began with a headache and nausea and later a high fever developed. The bilious character of the sickness which appears to be prevalent in all parts of Texas is indicated in the early stages by several symptoms, one of which is that the eyeballs turn yellow. In consequence of this unwelcome circumstance we decided to return to Torrey's Trading Post. When, on the following morning, we looked back on the charming little plain from the hilltop from which we had first seen the village, the thought involuntarily came to me how long it would remain in peaceful possession of the apparently harmless sons of nature, and if perhaps even now a land-hungry Yankee had cast his covetous looks upon it. My companion disillusioned me in a hurry, in that he assured me that not only this land, several days distant from the settlements, but also many other parts higher up on the Brazos, had been surveyed and had already been for some time the property of individuals. He also told me that the latter even now had the intention of asking the government to move the Indians to a more distant place.

Nothing worthy of note happened on our return journey, except that in addition to many buffaloes, we also saw two antelopes (*Antilope Americana Godman*) in the distance, and that while resting at noon for the purpose of cooking coffee, we set the prairie on fire accidently and were obliged to discontinue our siesta and beat a hasty retreat, so as not to be overcome by the smoke and heat. When looking back a little later, we noticed that the smoke had enveloped everything for miles.

We arrived at the trading post at noon on the second day. A continuous fever throughout the entire trip had weakened me to such an extent, that I was glad to find a temporary place of rest, even though it offered but a few comforts and I received but little attention. I was obliged to lie down immediately on my arrival and, semi-conscious from high fever and suffering particularly from the intense heat, I spent ten long, lonesome days in the log house. Quinine, calomel, and castor oil, the American panacea, which is found in every Texas home was given in large doses to allay the fever but without avail. My traveling companion, who had in the meantime returned from his trip on the Trinity, informed me that he was obliged to start back to New Braunfels on the following day. With considerable misgivings as to how to carry out the resolution, due to my weakened condition, I nevertheless informed my companion that I would accompany him, for I was convinced that if I should remain in this inhospitable place without care and medical attention, my sickness would end fatally. It was also probable that, if I did not avail myself of this opportunity, a long time would perhaps elapse before I should find a companion with whom to travel to New Braunfels.

Accordingly we started homeward on the following morning at daybreak. I was too weak to mount my mule by myself and had to be assisted by my companion. As a matter of fact, the auspices for a strenuous journey of several days through a desolate, uninhabited country could hardly have been imagined more unfavorable. However, after only a short ride, the fresh morning air had so revived me, that I felt better and stronger. My condition improved with every succeeding hour and for this reason we were able to travel twenty-six miles on the first day without exhausting me too much.

We stopped at Bucksnort during the noon hour. Here many people were also sick with fever which was not surprising, since the hamlet lay close to the broad bottom of the Brazos containing many sloughs full of stagnant waters constantly emitting mephitical odors. That the entire valley of the Brazos moreover is decidedly more unhealthily than the valleys of the western rivers as, for example, the Colorado or the Guadalupe, is no doubt due to the greater width of its forested, alluvial bottom.

The fact that my companion addressed me as doctor in the hotel where we stopped, gave occasion to an almost comical, yet embarrassing situation, for a well-dressed farmer asked me to accompany him to his home in order to administer aid to his very sick child suffering from convulsions. I explained to him that I was not a physician. This, however, received such little credence, since my companion had addressed me as doctor, that, assuming that I refused in order to avoid the inconvenience of a sick visit, the father of the child assured me that he did not desire my services gratis, but would pay me well. The poor man left me with loud lamentations and accusations for my indifference to help in such dire need, without paying attention to my further protestations.

After the heat of the day we covered the remaining six miles to the falls of the Brazos. The road led us diagonally across the Brazos bottom, which forms here a luxuriant virgin forest which was filled with half-rotted tree trunks strewn about in the wildest disorder.

The so-called falls are really only rapids and are caused by layers of bluish grey marlslate, belonging to the cretaceous formation, three to four feet high, lying diagonally across the river. The river was considerably broader here than downstream, but so shallow at this time of the year that we could ford it without danger.

We lodged in a house on the other side of the river. Three persons were sick here with malaria fever but the members of the family nevertheless assured us, according to the custom of the country, that the location of the settlement was healthful.

From the falls of the Brazos we took a direct road to Austin. The entire strip of land between these points has a uniform character. It is an elevated, gently rolling, almost treeless plain. The soil is everywhere fertile humus with a subsoil composed of white, less firm chalkmarl. Several crystal-clear brooks and rivers flow through this region. Since, in addition, the higher altitude of this region presupposes a more healthful climate than is found in the larger bottoms, and from the standpoint of agriculture there is hardly any other objection than that common in Texas—lack of wood—it is to be expected that within a short time numerous settlements will be made here. In fact, several farms were already established on the upper course of "Little River" during the past years. On this entire trip, lasting three days, we saw only three human habitations.

On the first day we rode thirty-five miles to the home of Mr. Bryant, a slave-owning farmer on Little River, whose settlement was already indicated several miles in advance by the numerous herds of cattle grazing in the prairie. Mr. Bryant complained to us that the Comanches had taken the greater part of his this year's half-ripe corn crop several days ago either by begging or by force. He expressed the hope that the government would reimburse him.

On the following morning we first had to cross the Little River. It is usually a clear stream, but was now swollen and muddy. We compelled our mounts to swim across, but we, ourselves, with our baggage and saddles crossed it in a canoe made of a tree trunk according to the custom of the country.

The Little River, although only an unimportant stream, has a bottom several miles wide. It took us several hours to pass through the deep, sticky mud. On leaving the bottom we had an immeasurable prairie, covered with mesquite trees, before us. After a ride of several hours we came to a little brook called Willes Creek where buffaloes in great numbers were grazing. They covered the grassy prairie separated into small groups and far distant on the horizon they were visible as black specks. The number of those clearly seen must have been not less than a thousand. When several hundred of them, disturbed by our presence, bolted, the ground resounded as at a charge of a regiment of cavalry. In their fright the herd came directly toward us and we had to hurry to get out of their reach, as they do not change the course of their flight, but trample under foot any object confronting them.

After a ride of forty miles we camped under the sky on the banks of the St. Gabriel River. This river is as clear and beautiful as the Guadalupe at New Braunfels, but of less volume. Despite the great natural advantages there were no settlements found here. Several years ago a German by the name of Schubert, who later was employed by the Verein, established a settlement a little lower on the river, but he had to abandon it on account of sickness and for other reasons.

On the following morning we rode fifteen miles before breakfast. This we ate at the isolated home of a Yankee on Brushy Creek who had lived here for two years. During breakfast our host informed us that he had in mind to abandon his farm and to move farther up the river. When my companion, who on a previous trip found this man well satisfied with his present home, asked in astonishment what induced him to make such a resolve, he answered in a tone of voice which sounded as though he were suffering from an intolerable condition: "The country is getting too crowded, I cannot live here any longer." This reason seemed peculiar to me since I had seen no houses far and near and I therefore asked him how close the nearest neighbor lived. "Well, the next fellow lives but ten miles from here," he answered.

I expressed my deep regrets to the man for being so hemmed in and at the same time thought to myself that it was, indeed, fortunate that everyone in old Europe did not require so much of the earth's surface as did this old backwoodsman.

From here it was only twenty miles to Austin, which place we reached at noon. During this ride we enjoyed the sight of the beautiful wooded hilltops among which the Colorado finds it source.

Taking the usually traveled route, we arrived at New Braunfels at noon, August 28, 1846. I greeted the sight of the Verein buildings, glittering in the sun, with joy, for several days prior to this I had had grave doubts whether I should ever see them again, owing to the state of my health. I also came to the conclusion at the end of the journey, that no place in the country, lying west of here can compare in beauty and other natural advantages with the German colony on the beautiful Comal River.

Chapter XVII
Electing Congressmen—Funeral & Bishop Odin— Mormons—Christmas—Insurrection at New Braunfels

Upon our arrival at the colony, we found conditions about the same as we left them. The shortage of funds in the Verein's treasury was as acute as ever. Just a few days prior, Herr von Meusebach had sent two mounted men to Galveston to inquire about an expected remittance from Europe. In the meantime Mr. Cabbes, a special authorized agent of the Verein, had arrived from Germany. He was to make a careful investigation as to the present conditions of the Verein and to find the cause of its past failures, in order to enable him to make a verbal report to the underwriters.

There were still many sick people in the city as well as in the country round about. Malaria and virulent fever were prevalent. None of the settlers along the Comal Creek were spared. I, myself, contracted malaria fever a few days after my return. The trip had not restored my health but only temporarily arrested the disease.

On September 7 several Lepan chiefs passed through New Braunfels who had just returned from the capital at Washington. In accordance with a custom followed by the president they had been presented with huge silver medallions, four inches in diameter. One side displayed a picture of the president, the other two clasped hands, symbolizing the friendship existing between the American people and the tribe of the Lepans. They showed these medallions with a great amount of satisfaction and in general seemed well satisfied with their visit to the capital.

The heat during the first half of September was just as great during the day as in August. From ten o'clock in the morning to three o'clock in the afternoon it registered between 90°F. and 95°F. However, the nights were already considerably cooler and the mornings foggy. No rain had fallen for several weeks. The grass in the prairie had dried, due to the long drought. The woods, however, were still resplendent in their verdant green.

During the night of September 25 the first norther appeared suddenly. After the long continued heat of the summer the sudden change was felt keenly. A heavy torrential rain accompanied the norther.

At this time several young Germans who had taken part in the war against Mexico returned, for their six months' term of enlistment had expired. Most of them returned dressed to a great extent in Mexican costume. All had acquired somehow one or more useful as well as beautiful woolen Mexican blankets. Some also brought with them as mementoes of this campaign highly ornamented Mexican saddles and bridles, as well as huge spurs with wheels as large as a dollar. One young, well-educated man, formerly from Berlin, who had been made adjutant to Governor Henderson since he had a knowledge of Spanish, told me much about the natural conditions of the country lying between the San Antonio River and the Rio Grande, as well as the valley of the latter river.

The San Miguel, the eastern tributary of the Nueces, establishes a conspicuous boundary on the road from San Antonio de Bexar to the Rio Grande, for on the other side of it no trees are found. There appear only thorny, dry shrubs forming at times impenetrable thickets called chaparrals, which caused the troop movement on the Rio Grande considerable trouble and often served the Mexicans as a convenient ambush. The cacti, which are in other parts of Texas known as shrubs, here grow to the height of a man. On the other side of the Nueces, the agave (*Agave Americana L.*) is found from which the national drink, pulque, is made and which is cultivated extensively in the interior of Mexico.

The valley of the Rio Grande in its lower course is in the main a monotonous region and does not impress one favorably. Near Matamoros it is a flat, alluvial plain in which corn, sugar and rice are cultivated. Farther upward there is less cultivated land and the farms are separated by infertile stretches. At Reynosa a solitary hill of solid rock, rises above the alluvial plain. Still farther up, between Camargo and Guerrero, whole ranges of hills of solid rock extend to the river's edge.

Wood suitable for building is so scare along the Rio Grande, that the lumber used in building a ferry boat had to be brought from Bastrop on the Colorado in the beginning of the war.

On October 3 a fierce norther made its appearance during the night and at the same time I became sick with dysentery. I had long feared this sickness, which had taken so many lives in the city and the surrounding country, because it usually follows in the wake of malaria fever, of which I again had several attacks. The illness appeared immediately in a virulent form and all remedies applied seemed to be without results. Just as my friends had given me up as lost and complete exhaustion had set in, I passed the crisis. However, a peculiar, feverish condition, which continued for a long time prevented me from regaining my strength.

On November 3 I had convalesced sufficiently to take a ride along the Guadalupe. Most of the trees had shed their leaves. The weather was extraordinarily clear and pleasant, as it is occasionally found in Germany during the first half of September.

In the bottom of the Guadalupe, hardly two miles below the city, we met a band of Delaware Indians who had been camping here for some time and who were hunting in the vicinity. They supplied us regularly with game and we paid them one bit, that is twelve and a half cents, for a large hind-quarter of venison.

During these days we also received very unfavorable reports concerning the state of health of the colonists at Fredericksburg. Fever and dysentery were also prevalent there, often terminating in dropsy which usually proved fatal. Out of five hundred persons, ninety-four had already died.

At this time the election of congressmen to Washington also took place which was the first to be held since annexation to the United States. The State of Texas had been divided into an eastern and western district for this purpose and each district was to elect a congressman. For the position in the western district two candidates announced, who in quick succession visited New Braunfels on an electioneering tour. One was a lawyer, the other a rich merchant from Galveston. Both tried their utmost, and with American adroitness, to prove their qualifications for the desired office and made the most definite promises, if elected, to secure for the German farmers of the west all possible advantages and rights, nor were they wanting in flattery for the "industrious and enterprising German citizens." The citizens, however, displayed very little interest in this election, although they had shown considerable interest in the previous ones for county officers. This may be explained by the fact that the people understood the significance and importance of a congressional election even less than the county elections.

On November 11, I was present at the burial of a young girl, the daughter of an agent of the Verein, Mr. K., who had recently come from Germany where he had been burgomaster at Anklam. A virulent fever had caused her death within forty-eight hours. According to a North American custom in the rural districts, all people in the funeral procession were mounted which appeared unusual as well as picturesque to a European. The burial did not take place in the local cemetery but, according to custom, on the property belonging to the father of the deceased. The place selected by the sorrowing father was beside the beautiful springs of the Comal in full view of the river and shaded by high forest trees.

At this time we also received a visit from the Catholic bishop of Texas, Monseigneur Odin, who wished to discuss with Herr von Meusebach the erection of a Catholic church in New Braunfels. He stated that it was his intention to build it with his own funds and he asked the Verein to give him the necessary ground for the erection of a building in the city free of charge. When this was granted, he stipulated very emphatically that the deed should be recorded in his name, evidently to prevent a possible refractory congregation from bringing pressure to bear in the erection of the building.

There were only two Catholic churches in Texas at this time. One was in Galveston and the other in San Antonio de Bexar. Bishop Odin, a Belgian by birth, as are so many of the Catholic dignitaries in the United States, lives in the style of the old gospel preachers, inasmuch as he travels continually about in the country, visiting the Catholics living scattered in the various parts of the country. Fearlessly and tirelessly he traverses the lonesome prairies on horseback, and through his restless energy and unassuming, charming personality has earned for himself the universal respect also of those not of his faith.

On the same day three elders of a Mormon colony, who had settled near Austin, came to Herr von Meusebach. They asked for permission to settle a company of forty-six families on the grant of the Verein colony.

After this peculiar, communistic, religious sect had been driven away by an enraged populace of Illinois from their former home in Nauvoo on the Mississippi, due to repeated robberies and murders committed by individual members, they had split into numerous groups which scattered in all directions. Such a group had come to Texas and had settled several months prior to this four miles above Austin on the Colorado. With great foresight and remarkable speed they had erected a mill on a brook emptying into the Colorado. This mill now produces most of the cornmeal used at Austin and New Braunfels and incidentally they have established for themselves a very profitable business. Several of my acquaintances who had visited this colony, were loud in their praises of the industry, order and frugal mode of living of these people. As a matter of fact I did not hear one word of complaint against them during my stay, on the contrary, their behavior was exemplary, although they were at first watched with suspicion. This company coming to Texas was therefore either the better element of the main body, or the

many accusations directed against the latter were caused by
a few black sheep in the fold, of whom they were later able
to dispose. This latter supposition seems plausible, for the
main body of this sect, after being driven out of Missouri,
moved west, intending to settle in California. However,
they finally settled on the rim of the Sierra Nevada, just recently explored by Fremont, and under such favorable auspices, the quick growth of a virile colony seems assured. The
three elders were not given an unqualified promise to their
petition; however, a contract was signed with them whereby
they agreed to build a mill at Fredericksburg similar to the
one they had built at Austin.

Since the middle of November the northers became more
frequent, appearing almost every other day. A fire of mesquite wood burning with a bright, smokeless flame in the
fireplace was now highly desirable as well as comfortable.
During the night of November 25 it froze for the first time.
Ice, as thick as a finger, still covered the ponds at nine o'clock
in the morning. The weather was very pleasant and warm
during the days when the northers were not blowing. The
thermometer registered 79°F. to 80°F. on these days. The
sudden change made the cold so disagreeable. A temperature of 55°F. seemed as cold to me as a temperature of 8°F.
at home in Germany. This same changeable weather continued throughout the month of December.

A few of the latter days were decidedly pleasant and compared with the beautiful October days at home. Thus on
one day the light summer clothes were comfortable, while
on the next not enough woolen blankets could be found to
keep warm.

According to a custom at home, Christmas Eve was
celebrated in company of the jolly companionship of the
Verein's officers around a richly decorated and illuminated
Christmas tree, for which a young cedar was used (*Juniperus
Virginiana L.*) The relatives and friends in the distant fatherland were remembered and many a one who had come
to Texas with high hopes and expectations may have now,

disillusioned, recalled with regret the comforts and joys of the native hearth which he had left so lightheartedly. Moreover the sight of the Christmas tree was calculated to arouse reflections on the rapid advance of the pale faces into the territory of the redskins. For on the same place where today the symbols of a happy German family life is planted in the midst of a cultured population, scarcely two years ago the camp fires of the wild Comanches were burning.

The last day of the year 1846 was to witness an insurrection in New Braunfels. In the early morning, numerous placards were fastened in various parts of the city through which the populace was asked to march en masse to Herr von Meusebach's home and compel him to fulfill immediately the promises made the colonists. In consequence of this, a mob numbering about one hundred-fifty persons armed with clubs and pistols came up the hill on which the buildings of the Verein stood. A deputation, composed of several individuals not enjoying the best reputation, went to the home of Herr von Meusebach. The rest contented themselves at first to wait for an answer from the delegation. When it was not forthcoming immediately, they crowded into the house and committed a number of excesses in the anteroom and uttered loud threats against the life of Herr von Meusebach. In the meantime the negotiations were carried on in the adjoining room. Mr. Henry Fischer, who had arrived from Houston a few days prior to this and from whom the Verein had bought the land, led the negotiations on the part of the deputation.

After deliberating for several hours, the following articles were agreed upon and signed by Herr von Meusebach:

1. All immigrants are to receive immediately the deeds to the one hundred sixty or three hundred twenty acres of land respectively, promised in the grant. Mr. H. Fischer is to safeguard the interests of the colonists in drawing up these deeds.

2. The surveying of the land apportioned each immigrant should be carried out without delay. Also in this

respect, Mr. Fischer was to take care of the interests of
the colonists, inasmuch as he was to see to it that the
surveying should be done at the proper time and in the
proper manner.

3. Several immigrants who did not receive a town lot
in New Braunfels, although they had immigrated here
under Prince Solms, should be given one.

4. Since the inhabitants of New Braunfels maintain
that the entire forest in the vicinity of the city was
promised them by Prince Solms and the managers of
the Verein, however, deny this, Prince Solms should be
asked by letter to settle this dispute.

5. Herr von Meusebach should hand in his resignation
to the directors of the Verein in Germany, but carry out
the functions of his office until his successor arrives.

Herr von Meusebach agreed readily to the latter provision, since prior to this he had pleaded with the directors to
relieve him of his office as general commissioner.

The result of these negotiations was announced by a member of the delegation and was received with loud acclaim by
the waiting mob. Thereupon they dispersed and went to the
city, there to proclaim their victory and to celebrate.

Thus ended this insurrection fortunately without blood-
shed, in whose origin one can distinguish between cause and
motive. In this particular case the real cause was no doubt the
dissatisfaction of the settlers over non-fulfillment of several
important promises made originally by the Verein. Most of
the settlers had been induced to join the Mainzer Immigration Society on the promise of a gift of one hundred sixty
acres for a single man, and three hundred twenty acres for
a man with a family. This promise was not kept upon their
arrival in Texas. The colonists furthermore were convinced
that the land intended for them, lay in an unknown, inacces-
sible mountain region, of which marauding and dangerous
Indians and not the society were sole owners.

The colonists had also ascertained that, if the individual
tracts of land were not surveyed by the coming August, the

right to the land on the Llano and San Saba would be for-
feited, according to a contract originally made between the
government of Texas and the promoters, Mr. Fischer and
Mr. Mueller. The prospect of losing all their claim would of
course also incite otherwise calm and decent immigrants to
take drastic action to safeguard them. The immediate motive for this insurrection was, however, the machinations of
a man who, to further his own selfish interests, was greatly
concerned in getting rid of Herr von Meusebach, and who
knew how to take advantage, in a sly and clever way, of the
personal enmity existing between a few individuals and
Herr von Meusebach.

Just how much the population of New Braunfels in general was in sympathy with the violent and unlawful manner
in which the petition was presented and especially with
the unfair and undeserved treatment accorded Herr von
Meusebach showed itself the next day in a mass meeting,
where, in addition to other persons, most of the Americans
living in the city, who were capable of forming an impartial
judgment since they were not directly involved in the quar-
rel, were present. At this meeting the methods employed
were condemned in strong terms and the chief justice, Mr.
Dooley, emphasized with indignation how lawless it is to
violate the sanctity of the home.

The insurrection did not bring about any changes in the
Verein's management, since Herr von Meusebach continued his duties until the arrival of his successor. Herr von
Meusebach now energetically pushed the preparation for
an expedition to the land north of the Llano, which had
been long planned but always delayed on account of more
important business. Through this expedition an end was to
be made to the uncertainty as to the worth of these lands, for,
as implausible as it may seem, neither the vendors, Messers.
Fischer and Mueller, nor agents of the Mainzer-Verein had
ever seen it and when thousands of German immigrants,
who were to settle in this region, were already in Texas. The
trip was generally considered very dangerous on account of

attacks by the Comanche Indians who dwell there. To assure
a measure of protection, the number of participants in this
expedition was brought to forty men. Most of these were
engaged with pay by the Verein for the duration of the trip.
In addition to this several young men joined as volunteers
to satisfy their curiosity or for the love of adventure. A visit
to the San Saba Valley, the unknown, almost mythical won-
derland with which every Texas settler associated the idea of
unsurpassed fertility and loveliness, and at the same time a
wealth of precious metal, promised to satisfy both.

Included in the preparations for this trip was the buying
of various presents for the Indians, so-called Indian goods,
consisting primarily of woolen and cotton goods. Whenever the Indians and whites meet on a mission of peace, the
former always presuppose that the latter must give presents
and they make this a kind of stipulation for further peaceful meetings. They argue from the premises that the whites
have been favored with more wealth than the red man by
the great spirit, and therefore have a moral obligation to
share their riches with them. These presents dare not by any
means be worthless trifles, as is erroneously supposed in
Europe but the Indians know full well which are good and
genuine articles of practical value and not mere playthings.
A present to an Indian chief for example will cost the donor
not less than $15.00.

The company, divided into several groups, left New
Braunfels January 14, 1847. The final provisioning was to
take place in Fredericksburg from where the actual expedition was to start. I, myself, saw them depart with a feeling of
painful resignation. For a long time I had waited for such an
opportunity to become acquainted through personal observation with that unknown hilly country of Texas, on whose
soil no naturalist had ever set foot, and in which, according
to a generally accepted supposition, a higher chain of hills
arises where interesting geological conditions are found.

But at the time of the departure of the expedition I again
felt indisposed. I really had not recovered sufficiently from

the virulent climatic sickness to dare to go on a trip lasting several weeks, and which evidently would be connected with great hardships and exertion. It was the more difficult for me to bow to the inevitable, as it was very probable that during the rest of my stay in Texas no other opportunity would present itself to see the land of the Indians on the San Saba.

Moreover, the departure of the expedition fell during the coldest part of the year. The northers were more frequent and colder than in December. During the night of January 8, the water in my bedroom froze the thickness of a finger and at 8:30 in the morning the thermometer registered 22°F. The cold seemed more noticeable to me than anytime in Germany. Four woolen blankets would hardly keep one warm in bed.

However, on the following day the thermometer registered 82°F. in the shade at two o'clock in the afternoon. Similar extremes in temperature were recorded several times during the following days. The weather was generally mild and pleasant on all days in which there were no northers.

Chapter XVIII
FREDERICKSBURG—SURVEYORS—SHAWNEE INDIANS—DYSENTERY
ARRIVAL OF INDIAN AGENT—MEUSEBACH EXPEDITION

On January 20, I traveled to Fredericksburg in company with an agent of the Verein, for the state of my health had improved sufficiently in the meantime to undertake such a short trip. Two wagons and six people accompanying us prevented us from traveling as rapidly as on horseback, however, we had the advantage among other things of being able to carry with us buffalo robes and woolen blankets which afforded us complete protection at night against the northers. This slow mode of travel also gave me an opportunity to observe the natural conditions of the country more leisurely.

Fredericksburg lies in a northwesterly direction from New Braunfels; however, the only road passable for wagons does not follow this course, but describes a considerable curve. At first it runs in a southwesterly direction as far as the Cibolo where it forms a juncture with the San Antonio Road. From this point it takes a straight northwesterly course, following an old Indian trail, called the Pinto Trail, selected with Indian sagacity, which is the only convenient passage from the undulating prairies of western Texas to the interior of the rocky tableland, cut up by deep ravines and fissured river valleys.

We left in the afternoon and traveled only eight miles on the first day of our journey. An icy cold norther was blowing when we pitched our camp under some live oak trees near a waterhole. The following day brought us to the dry limestone bed of the Cibolo. Two miles beyond this river the road branched off. The one to the left went to San Antonio, the one to the right to the endless wilderness in which, at that time, for a distance of ninety miles to Fredericksburg, there were no human habitations.

The road leading into the wilderness was, however, at this time, indicated more distinctly and clearly than the San

Antonio Road, since recently a number of loaded wagons had gone to Fredericksburg frequently. Just beyond where the two roads branched off, the road began to climb and the soil assumed the stony character of the hilly country north of New Braunfels. However, at no place were the slopes as steep and precipitous toward the undulating land as there. Sparse forests of post oaks and isolated live oaks predominated here. Toward the north, extending as far as the eye can see, rectilinear, barren hills arose above each other which, with their dry, yellow grass covering, conjured up impressions of desolate loneliness.

We camped in a little valley, about twenty-two miles distant from New Braunfels. This was the first point on the other side of the Cibolo where water is found, and which therefore served as a regular station for the wagons going to Fredericksburg. It had received the rather uninviting appellation of "mudhole." In fact the water supply consisted of a small, dirty puddle from which hardly a few buckets of muddy water could be obtained. In this respect one is, however, not so particular while traveling in Texas, for if it is not too bad for the draft animals, the traveler helps himself by imbibing the fluid in the form of coffee, in which the dirt is less noticeable. That there was no other water for miles in this region was attested by the fact that several times herds of deer came to this puddle and our presence scarcely drove them away. My companion was fortunate enough to kill two deer and thus we had a sufficient supply of fresh meat the rest of our journey.

During the night a prairie fire caused us considerable worry since it approached us from several directions. We resorted to the usual method of protection by burning the grass for a certain distance around our camp. Not until we had thus assured our safety, could we enjoy the beautiful spectacle.

In the darkness of the night the strips of fire, several miles in extent, appeared as fiery bands, which, governed by the strength of the wind, moved forward now quickly, now slowly; the flames shooting up high or just glimmering, according to the length of the grass. An especially beautiful view was afforded by a group of live oaks, on which magical illuminations were cast by the burning grass.

With reference to the dreadfulness and danger of such prairie fires in Texas, many exaggerated reports have been spread, especially have these been stimulated by a bombastic, romantic description in a book, referred to in the introduction: *Mons. Violet's Adventures in Texas, etc.*, which in ridiculous exaggerations is hardly surpassed by other passages of the book, rich in fabrications. As a general rule the fire does not advance so quickly that one cannot escape with ease. Furthermore, the burning strip is usually so narrow that one could easily jump over it. Only when fanned by heavy winds and when the grass is exceptionally long, do the flames move rapidly and the heat become intense. As a rule the fire is not fierce enough to consume shrubs of the thickness of a finger which lie in its path, but it merely destroys the leaves. However, in such an event a mounted man would have little difficulty to escape from the approaching fire or to find a place where the grass is short. These fires often cause severe losses near the settlements, since they destroy the fences enclosing the fields. When the settlers sight a prairie fire nearing their farms, especially during a strong wind, all the inhabitants hurry to the threatened place, and an effort is made to extinguish the flames by beating them with wet sacks, or a ditch is drawn to halt their advance.

Our road on the following day led us at first over similar rocky heights covered with oaks. Before reaching the valley of the Salado, we met a company of about twenty men who had been sent by the Verein to repair the road between New Braunfels and Fredericksburg.

The valley of the Salado has here no constantly flowing streams, but only a few waterholes. These dry beds in so many valleys of the hilly parts of Texas become raging torrents during heavy downpours in the spring, similar to those of southern Europe carrying with them rubble and tree trunks, while during the rest of the season the riverbed in the center of the valley is dry.

At this point our road in the valley of the Salado led us to higher elevations. The soil of this rather narrow bottom is excellent and a few farms could be advantageously established here, where the almost total absence of wood would be the only disadvantage. In fact several miles above this point, where the road from New Braunfels runs into the valley, near a spring called Comanche Springs, Herr von Meusebach, who had resigned his position as manager of the Verein, has settled since my return to Germany where he intends to cultivate his field in rustic seclusion.

On the hillslopes of both sides of the valley, a peculiar plant, a species of the genus *Dasylirion* with rush-like, sublated leaves, unknown in the low parts of Texas, forms large bright green patches, which contrast sharply with the surrounding dry, yellow grass. With the exception of this plant, only the stemless yucca with upturned leaves, displayed any green foliage during this time of the year.

After a day's journey of twenty-four miles we camped near several springs which were designated by the simple appellation "The Hole," and which formed a regular station for the German teamsters going to Fredericksburg. We met unexpected company here. In the bushes surrounding the springs, a number of American surveyors were encamped, who were employed by von Meusebach to survey the grant and who were now on their way to Fredericksburg. All were wild looking fellows, dressed almost entirely in deerskin and armed with a long American rifle and the bowie knife in their belts.

Surveying in the uninhabited highlands of Texas is not the peaceful, dangerless occupation as in Germany, but always connected with danger and great hardships and privations. Camping under the blue sky for months, often many days' journey removed from the homes of civilized people, the Texas surveyor finds his rifle just as necessary as his compass, on the one hand to supply meat for his needs, on the other to ward off attacks by hostile Indians. The latter, who regard the compass as the instrument or "thing which steals the land," know full well that the surveyor is only the forerunner of the white intruder who will drive them off the hunting grounds of their fathers. Therefore they pursue him with particular hatred. Although surveyors venture into such a country only in companies of not less than six to twelve men, it nevertheless happens every year that such companies are attacked by Indians and all or some of the men are killed. I recall such a case during my stay in Texas, where three of a company of eleven men were murdered at the springs of the Rio Blanco.

When a private person acquires a piece of unsurveyed land from the government, either for services rendered during the war or through purchase, he asks the district surveyor, maintained by the State, to survey the land. This officer then commissions one or more deputies to do the work, who in turn engage as many mounted armed men as are necessary for protection against the Indians. A specified tax, based upon the number of acres to be surveyed, defrays the expense of this work.

On the following day our road led us again over barren, sterile limestone hills, where the horizontal beds of hard cretaceous limestone were exposed everywhere and no soil suitable for farming was seen. We had to cross several small creeks with narrow bottoms and steep rocky approaches. The first was the Sabine about eight miles from "The Hole," and about four miles farther was Wasp Creek. This side of Sabine Creek, we found great quantities of fossils in the white chalkmarl of a ravine through which we passed.

Several miles farther we descended into the Guadalupe valley, which presented a most pleasant contrast to the barren, stony highland. The rather wide valley on both banks of the Guadalupe had the appearance of a most beautiful natural park. Everything tends to make this an excellent place for several farms. Several rows of very old cypress trees six feet in diameter, resembling a row of mighty columns, stood near the ford of the river which was clearer than that at New Braunfels, but contained less water. The river can be crossed regularly with wagons here at all times without danger. Only after a heavy downpour does it occasionally swell temporarily to such an extent that passage is impossible.

After a journey of twenty-four miles, we selected our camp four miles the other side of the ford at a little creek, called Cypress Creek, which empties into the Guadalupe. A number of wild turkeys flew into the cypress trees at dusk and we were fortunate enough to kill two, thereby supplementing our meat supply in a welcome way. We covered the remaining thirty miles to Fredericksburg the following day, enjoying the most beautiful, clear weather which to us seemed almost too warm for the season.

On leaving our camping place, we entered a beautiful, grass covered valley extending several miles upward which was covered with numerous clear creeks called "Sister Creeks." Toward the north we saw the summit of a high hill which the German immigrants named Sarberg because of the similarity of outline. Farther on, the valley becomes more narrow and the bottom begins to rise sharply. The scattered live oaks here are dwarfed and by their stunted growth indicate the considerable elevation in which one finds them. Upon finally reaching the last heights, it becomes apparent that one now is on a high, narrow plateau. The ridges of the highest surrounding range of hills are on the same level with this plateau.

This is the divide between the Guadalupe and the Colorado, or rather the Pedernales. The elevation at this point is hardly more than two thousand feet above sea level. Looking toward the valley from which we had ascended and on the barren, rocky summit of hills which arise as far as the eye can see to the west, east and south, the view appears wild and romantic, leaving the impression of original primitiveness, since there was no sign of human activity or human habitations. If someone who had only seen the more accessible lowlands of the coast, and the gently undulating prairies of the hilly country were suddenly transplanted here in the rough mountainous country, almost Alpine in character, he would never recognize it as a part of Texas.

Moreover, this highest elevation between New Braunfels and Fredericksburg is formed of horizontal, alternating soft and hard beds of white limestone, belonging to the cretaceous formation, as is generally found in the whole highland between these two points. Here at the highest point are found, at several natural exposed places of the rocks, a number of fossils, particularly a great number of *Exogyra Texana*, which are distributed from New Braunfels to the San Saba River.

After traveling a short distance in a grove of stunted oaks, we came to the northern descent of the plateau. Here also a panoramic view met our gaze. A wooded valley, several miles wide, was visible in the distance and beyond it a barren ridge of hills was outlined against the horizon. This was the broad valley of the Pedernales to which our road now gradually descended. Before coming to the river we passed through a beautiful mesquite prairie of great fertility. Here we were surprised to see a large herd of cattle, belonging to the colonists at Fredericksburg. They had been driven here since there was not sufficient pasturage for them in winter near the city. The Pedernales itself is a small, rapidly flowing stream with considerably less volume of water than the Guadalupe, but of equal clearness. An important bottom, in which tall pecan trees (*Carya olivaeformis Nutt.*) are frequently found, borders its banks. The oak forest begins immediately on the other side of the river and extends even beyond Fredericksburg and for many miles along the river. There is at the same time a change in the composition of the soil for here it is light and sandy, which no doubt accounts for the prevalence of post oaks (*Quercus obtusiloba Mx.*) which prefer a dry, sandy soil.

The remaining six miles from the ford to Fredericksburg were covered quickly, as our horses and mules traveled more lively, evidently sensing the end of our several days' journey. We came to the first house at sundown. It took a little more than four days to make this journey of ninety miles. As regards the condition of the road from New Braunfels to Fredericksburg, it is in general as good as any other similar road in Texas of equal distance, now that the improvements instituted by the Verein have been carried out. Even if there are a few rough places and several steep inclines, it is entirely free from swampy stretches and muddy river crossings, which are so prevalent in the lowlands of Texas and cause the wagons so much trouble and delay. Even the heavy spring rains do no appreciable damage to the road, since its base in the main consists of solid layers of rock. In general, horses or even more so mules, are preferred for drawing the wagons, as oxen, usually used as draft animals in Texas, cannot get along so well on these hard, rocky roads.

After sighting the first house, we still had to cross a little sluggish stream and shortly thereafter we entered the long, principal street of the city. The Verein's building where we halted was one of the first buildings of this street. It was a rather rough, one-story log house which served as living quarters for the agents. An adjoining room was used as a warehouse. A palisade constructed of strong posts driven into the ground enclosed the yard in which horses and mules could be kept to prevent them from straying or being stolen.

The entire yard was built in a manner to withstand successfully an Indian attack. Such attacks were by no means improbable at the time of the founding of Fredericksburg and it is remarkable that the builders of the city suffered so little from the hostility of the Indians. There was a little cannon in the yard, also intended for defense against the Indians, but fortunately it was never used.

We found Herr von Meusebach still here. The greater portion of the men of the expedition were already encamped nine miles in the direction of the Llano, and he, himself, was ready to leave in the morning, as some final arrangements had caused his delay. On his trip from New Braunfels he had lost three of his best horses, and it was never ascertained whether they had strayed or had been stolen by the Indians.

On the following morning after Herr von Meusebach had left, accompanied by our best wishes for the success of his hazardous journey, I decided to become better acquainted with the city and its vicinity.

Fredericksburg is situated on a gently rising plain about six miles north of the Pedernales Creek, between two small creeks which form a juncture immediately below the city. The western one, Meusebach Creek, is the larger of the two. A dense, uniform oak forest covered the area on which the houses were now being erected. This forest extended over almost the entire surrounding country with the exception of a small strip of open prairie, which ran parallel with the larger creek. The stumps in the streets were by no means all removed. The city lies southeast by northwest which is also the direction of the principal street about two miles in length. The main street, however, did not consist of a continuous row of houses, but of about fifty houses and huts, spaced long distances apart on both sides of the street. Most of the houses were log houses for which the straight trunks of the oak trees growing round about furnished excellent building material.

Most of the settlers, however, were not in possession of such homes, since they required so much labor, so many lived in huts, consisting of poles rammed into the ground. The crevices between the poles were filled with clay and moss, while the roof was covered with dry grass. Some even lived in linen tents which proved very inadequate during these winter months.

When following the main street, one comes to the market square which appears to be large enough to accommodate a city of ten to twelve thousand inhabitants, but which at this time was still covered with trees. Several laborers were, however, engaged in felling trees. On the other side of the square, a little log house was being erected which was a humble beginning of the beautiful public and private buildings which the founders of the city visualized. Very few houses and huts indicated the direction of the street on the other side of the market square and it finally vanished in the forest.

From this description one would conjecture that the place does not deserve the name city when measured by our European standard. However, in America everyone is liberal with such designations, since the near future is considered the present. If the location is well chosen, a few years will often justify the name city. On the other hand, some settlements do not progress beyond the first few log huts or finally disappear altogether, as the traveler in the western states often experiences to his annoyance. Often when he arrives at the place where his road map indicates a city or a hamlet in large letters and he intends to lodge there for the night, he finds only two or three dilapidated abandoned log houses.

The soil at Fredericksburg is light and sandy and the subsoil is composed of brick red, tough clay, which is found at a depth of one and one-half feet. Along the larger creek, in the open prairie previously mentioned, is a heavier soil rich in humus. The soil, generally speaking, does not compare favorably with that of New Braunfels or some sections in the lowlands. The sparse grass and the post oaks indicate soil of less fertility. In seasons when it is not too dry, good crops of corn and other grain may, however, be raised.

It would have been much more advantageous if the city could have been built on the Pedernales itself, but the land there was owned by private persons and could only have been purchased at a comparatively high price. The number of inhabitants in Fredericksburg at the time of my arrival was six hundred, but it increased to nine hundred within a few weeks. Every family and every adult male was assigned a one-half acre town lot and additional land near the city was promised each one. The tract of land at Fredericksburg owned by the Verein contained ten thousand acres.

In the afternoon of the same day I rode to one of the hills lying about two miles north of the city to get a better view of the country. This was not possible from the city itself, since it was encircled by a forest. Many logging roads crossed the forest and robbed it of the primitive appearance which it had only a few months ago. Moreover, nothing is farther removed from the European idea of a virgin forest than a Texas oak forest, even when in a state of primitiveness or untouched by human hands. In it are no trees of great height or huge thickness, and their height is seldom more than thirty to forty feet, and the diameter one and one-half feet. No vines climb up on them or form festoons in their tops; no impenetrable thickets cover the ground and one can walk unhindered among the trees, since all underbrush is lacking. These forests, however, are similar to our cultivated oak groves, since they contain only oaks of one species, whereas other American forests distinguish themselves from them through the manifold varieties of trees.

Moreover, the forest had a very desolate appearance, not only because the trees were bare, but also because the ground was covered with black ashes. Through the negligence of several settlers, the grass had caught fire in the neighborhood of the city and had spread several miles beyond it.

Not far distant, I saw an animal the size of a mink, with long, bushy tail and a broad white stripe running down its back, slowly running about on this black soil. Every doubt as to what it was, was soon removed as I approached it more closely, for I was met by a violent penetrating odor somewhat resembling burnt horn. It was the polecat or skunk (*Mephitis putorius Gmel.*) of which, as I learned later, many lived in the neighborhood of Fredericksburg. They find a welcome shelter in the cavities at the foot of trees, caused by the frequent forest fires. In such a hole this one also found refuge just as I was about to overtake it. They move very slowly and one can easily overtake them, as the front feet are adapted primarily for digging, which can readily be seen from their manner of walking. They stand still just at the moment when they squirt a greenish, yellow, stinking fluid (manufactured by two glands near the anus) ten or twelve paces. While doing so, they hold the tail high up in the air. The numerous specimens which I have seen at Fredericksburg and other parts of Texas varied greatly in the shading of the pelt and the distribution of the white stripe, which caused them to be classed erroneously as several species.

On the slope of the round hill, one hundred feet high, where I had decided to go, lay several bleached buffalo skulls. Since the settlement had been established, the buffalo had withdrawn from this region. From the summit of the hill one could see an extensive forested area and also get a glimpse of the scattered houses of the city. The view to the south was obstructed by a chain of hills which form the divide between the Guadalupe and the Pedernales. In the other directions, round, barren hills, similar to the one on which I was standing, form a semi-circle around the area in which the city lay.

All these hills consist of horizontal beds of white marl and limestone belonging to the cretaceous formation and containing numerous chert globules. In the clefts or other exposed places in the rock were found numerous fossils which do not leave any doubt as to the age of these layers.

Collecting these fossils occupied most of my time while in Fredericksburg. The many new and unknown forms of a prehistoric fauna in a virgin territory, not trodden by a paleontological predecessor, had a charm for me which only a naturalist can really appreciate. Upon returning to the city, I found a number of mounted Indians with pack horses in front of the Verein building. They were Shawnees who had been camping in the vicinity of Fredericksburg for some time for the purpose of hunting. They brought with them bear meat and particularly bear fat, which they offered for sale. The fat was in liquid form and clear. It was kept in deer skins, which formed a simple, but serviceable vessel. The entire deer skin is used after the hairs have been scraped off and the openings in the legs and the neck have been tied securely.

A gallon of this fat or bear oil was sold for a dollar by the Indians or for the equivalent in goods. The colonists came from all directions with all manner of vessels to buy, since this oil was suited for various purposes in the home in place of lard or oil. Not only was all the food cooked with bear oil during my stay in Fredericksburg, but the colonists also used it in their lamps instead of regular oil. Since it is easily digested, there was an advantage in using it in the preparation of foods instead of other fats. How plentiful bears were near Fredericksburg was attested by the fact that each Indian often had sixty gallons of such fat for sale. I found bear meat very palatable, resembling pork.

There were several tall, stately men among the Shawnees with well formed features, however, the majority were small with broad, typical Indian faces. Some of them spoke enough English so that we could easily converse with them. After the bartering over the bear meat and oil was satisfactorily concluded, the chief asked Dr. S., the Verein manager, for a written testimony as to the good behavior of his band, so he could show it to other whites when he met them. After receiving it, he rode off with his men with many protestations of friendship.

During the night a large gray wolf (*Canis occidentalis Richardson*, referred to as "loafer" by the American who shot it) was killed by a settler. This wolf was so bold as to come up to the house to steal a piece of meat. Such wolves were found frequently near the city, but proved harmless to human beings here as well as in other parts of Texas.

On the following day I rode to a place near the city where the majority of the male settlers were engaged in felling trees. Since most of the colonists were not even finished with the building of their homes, it was foreseen that they would not be able to enclose their fields with fences in time to plant corn. Raising corn was a matter of life and death, since upon it depended the existence of the colony. Upon recommendation by the manager of the Verein, all had agreed to enclose several hundred acres with a fence, and each family would then be assigned a proportionate share of the land for the raising of corn. The sound of the axe and the crashing of falling trees could be heard in all directions. The straight trunks of the oak trees were split into fence rails; the limbs and twigs were gathered into piles and burned. Everyone worked industriously and seemed to feel that the future success of the colony depended upon this work. After returning to Europe, I heard to my great joy, that this industry and toil, through which the colonists proved themselves real Germans, was rewarded with a good corn crop.

On the following morning, January 28, everything was covered with frost. The heat, however, became almost oppressive during the day. These sudden changes in temperature occurred quite often. Not infrequently, a real winter day would be followed by warm weather which could be compared to a clear, beautiful May day in Germany.

Unfortunately the sudden changing weather condition seemed to be just as harmful to the state of health as the extreme heat of summer. Dysentery and stomacace were still prevalent among the colonists. The latter, a loathsome and protracted disease, almost unknown in Texas and occurring seldom in New Braunfels, often proved fatal, since it spread to the throat and lungs. Almost every day one or more deaths occurred. Medicine and medical attention were furnished without cost by the Verein. It was a pitiful sight to see men, women and children gather daily with bottles at the room which contained the drugs, at a designated hour. Some of them were so weak and sick themselves that they could scarcely return home, but they were obliged to do so, since the other members of the family were bedfast. Sometimes the neighbors would carry the medicine because everyone in the house was sick. In consequence of the heavy demand, the supply of some of the medicine became exhausted, particularly citric acid which was used to combat stomacace. During the time that new supplies were brought from New Braunfels, which took several weeks, substitutes had to be used which were not as effective.

The cause of the prevalence of this disease, which ordinarily is not found in this country, can be traced very likely to the lack of a variety of food. For months the immigrants had subsisted almost entirely on cornmeal, beef and coffee and there was particularly an absence of fresh vegetables. It may be that the location of the place, enclosed as it was on all sides by woods, where the wind had little access, was the cause of the spread of this disease.

Since May 21, 1846, when the first contingent of colonists arrived here, the deaths numbered one hundred fifty-six, which is exceptionally high in proportion to the six hundred souls forming the entire population. The death rate was highest during the hot summer months and at that time shocking incidents of human misery occurred. An agent of the Verein told me how he discovered the half-decomposed bodies of a man and his wife in their hut, where they had died unknown to anyone and without receiving help. On another occasion, a father and his three children were taken to the cemetery on one day in the same cart. Here, also, the misery was accompanied by brutality and demoralization, as was the case when the immigrants came from the coast to New Braunfels. Oftentimes the dying were robbed by their neighbors and the estate of the deceased was considered common property which anyone could appropriate.

On February 5, Major Neighbors, agent of the United States Government of Indian Affairs, arrived unexpectedly from Austin with three companions. He was the bearer of a letter from the Governor of Texas to Herr von Meusebach, in which the latter was dissuaded from carrying out his plan of going into the Indian country, of which plan the Governor had heard. His attention was particularly called to the dangerous consequences, in view of the war with Mexico, should the expedition arouse the hostility of the Comanches against the outlying settlements.

In the event that Herr von Meusebach had already started on this expedition, Mr. Neighbors had orders to overtake him and to offer his services and assistance in negotiating with the Indians, as he was personally acquainted with several of the Comanche chiefs, and also had an intimate knowledge of their customs and habits. In the latter event, one of his companions, a half-civilized Indian chief of the Delaware tribe, Jim Shaw by name, who understood the language of the Comanches and also knew English, having lived a long time among the whites, was to act as intermediary and interpreter. When informed that the expedition had left ten days ago, Mr. Neighbors decided to overtake them without delay.

As my condition had improved in the meantime, I resolved to make use of this opportunity to see the unknown Indian land on the Llano and the San Saba rivers. My preparations were of the simplest kind and were completed within a few hours. I selected a large, fast mule which was also to carry my baggage. The latter was confined to my rifle, several pistols, several pounds of flour, ground coffee, sugar and salt, and a tin drinking vessel, holding about a quart, which was fastened to the saddle. This cup could be used for dipping water as well as for cooking coffee. I depended for a regular supply of meat upon the experience of my companions as hunters. Two woolen blankets, one under the saddle and the other on top of it, were to serve as a bed and cover during the night.

Chapter XIX
The Llano River—San Saba Valley— Comanche Indian Camp

On February 6, equipped in the manner described, I left Fredericksburg toward evening and found my companions encamped about four miles northwest of the city. Since the grass had been burned everywhere in the vicinity of Fredericksburg, they had hurried to this place to find some for their horses.

The camp, pitched under several live oak trees near a clear brook, presented a friendly appearance, although it was winter time when all the trees were bare and the grass in the prairie yellow and dry.

On February 7, we arose at sunrise and after a short delay, caused by the preparation of our breakfast consisting of coffee, fried bacon and bread, our little company was on its way. Jim Shaw, a six-foot-tall, strong Delaware chief, led the way on a beautiful American horse. Viewed from the rear, he looked quite civilized, since he wore a dark, stylish cloth coat which he had bought in Austin in a haberdashery, and a black semi-military oilcloth cap. Viewed from in front, his brown features, however, betrayed his Indian origin immediately; and upon closer examination one found that his European dress was by no means as complete as it appeared, for it lacked what is generally assumed to be a very essential part of a gentleman's dress, namely, the trousers. Instead of these he wore deerskin leggins, similar to our riding leggins, which reached halfway up his thigh. Then followed Mr. Neighbors and I, then a young American whom Mr. Neighbors had engaged for the duration of the expedition, and a common Shawnee Indian. Each of the two latter drove two pack mules which belonged to Mr. Neighbors and Jim Shaw. They were loaded with various kinds of articles of trade, particularly woolen blankets and cotton goods, for which they wanted to trade mules from the Comanches.

The Shawnee was an ugly little fellow with broad, high cheekbones and a dull, bestial expression in his face. He rode a little Indian pony, a malicious, tricky beast, as most Indian ponies are, which one could not approach within a few steps with impunity, and only its master could manage it. Whenever he spurred the stubborn mules on to a quicker pace with rough, inarticulate sounds, galloping through the bushes impetuously with hair flying about, one was in doubt which of the two, horse or rider, surpassed the other in untamed ferocity.

Our road, or rather wagon track which we followed, made by the expedition preceding us, led us at first through a valley covered with live oaks and post oaks. Later we came to a stony, infertile plateau, which, on account of the stunted oaks and exposed limestone visible in many places did not present a very cheerful view and it seemed all the more cheerless since all the grass had been burned as far as the eye could see. Only once was this cheerless picture broken and that in an agreeable manner by a little valley, in which several springs, tributaries of the Pedernales, flowed over solid layers of limestone. Toward evening we descended from the stony heights into a broad valley covered with a rich growth of grass and scattered mesquite trees, always the sure sign of fertility, which appeared as an oasis in the desert. We found fresh buffalo tracks here, without, however, sighting any buffaloes. We decided to spend the night near a spring where a gourd and other signs indicated that some one had camped here previously. During the course of the day we had passed two other camps of the von Meusebach expedition, since we traveled much more rapidly than they did with their heavily laden wagon, which hindered quick travel over this roadless wilderness. We had now covered about twenty-six miles, traveling in a northwest direction from Fredericksburg.

By February 8, we had gone only twelve miles in the direction of the Llano. Our road again led us for the greater part over stony, sterile heights. Due to the fire, all the grass had also been burned here, and only a few stemless yuccas and a few cacti of the species *Mammillaria*, growing in the clefts of the rocks, were spared. A forest, extending along its bank, announced to us that we were nearing the river. The Llano is a broad stream and of greater volume of water than the Guadalupe at New Braunfels. Like all rivers of the rocky tableland of Texas, it flows rapidly and is as clear as a crystal. Before reaching the river itself we had to cross a broad gravel bed which, together with the fact that we saw cane lodged in the trees, was a sure sign that the Llano like all other streams in Texas, is subject to considerable annual overflows. On both banks of the river and especially on this side there was a narrow strip of fertile land, which, as the fresh signs on the trees indicated, had just been surveyed by the men whom Herr von Meusebach carried with him, thereby securing it for the Verein for the protection of German immigrants. Here were located the previously purchased so-called head-rights, i.e., land grants from the State for war services rendered.

We decided to camp here for the rest of the day, although it was only noon, since there was tender green grass under the trees of the forested banks of the river which had been protected from the night frost and now promised our exhausted animals welcome food. It was also necessary for us to stop here in order to replenish our meat supply through the hunt which is always more successful along the streams where game prefers to stay during the winter time, due to good pasturage. We soon accomplished our purpose, for Jim Shaw, our Indian companion, bagged a deer and a turkey and Mr. Neighbors caught a number of catfish (*Pimelodus*) each a foot or more long. The craw of the turkey, with the exception of a few acorns, was filled with a great number of leaves of a kind of leek (*Allium*) growing wild here in abundance, which gave the flesh an agreeable onion-like taste.

On February 9, we crossed the Llano and advanced thirty miles in a northwesterly direction. The ford was only three feet deep and could be crossed conveniently. Crossing the river, moreover, had a two-fold significance for us, for we had now reached the southern boundary of the grant belonging to the Verein and with it the actual territory of the Comanches. The character of the soil changed immediately upon reaching the other side. Instead of the gravelly limestone of the cretaceous formation which forms the predominant rock from Fredericksburg to this point, there comes in a brown, ferruginous, evidently metamorphosed sandstone, whose weathering produces a light, loose soil. A few miles farther, the character of the region changed again and the sandstone was replaced by a red granite, which arose in a number of places in various shaped erect formations, but reaching nowhere any great heights. These formations often stood so close to each other that there remained scarcely a narrow opening for the Indian trail which we were following. The land in general, however, appears by no means to be sterile, for between these granite elevations are sizeable plains, overgrown with oaks and mesquite trees, and the loose soil, composed of decomposed granite is without doubt suited for agriculture. Several clear brooks, whose beds contain great masses of granite, worn smooth by the action of the water and whose banks are lined by willows, seldom seen in Texas, water this strip of land.

As well as I could determine while riding along, the granite in most places seemed to be of coarse grain and red from the color of its felspar. Huge veins of white quartz cut the granite. After traveling twenty miles, we crossed a ridge, running east and west, composed of steep, ascending limestone beds, but granite appeared again immediately on the other side.

We camped at a place where we found some dry, long grass which furnished our horses meager food. Due to the fact that the place was surrounded by a brook like an island, the grass had been spared by the general fire.

We left early in the morning on February 10, after a severe frost and while an icy north wind was blowing. We entered a post oak forest which, as we later noticed, extended a number of miles and which formed the only larger, continuous forest we had seen thus far on our trip. Several times we had difficulty to follow the trail of the von Meusebach expedition preceding us, since the hard soil barely left an imprint, and the wind had obliterated it still more by covering it with ashes. But our Indian guide always overcame this difficulty with his keen sight. On this as well as on many other occasions I had ample opportunity to convince myself that the Indians really possess such a keen sight and that this claim is by no means an exaggeration, but it should be mentioned that also the whites develop a similar keen sense when leading a life such as the Indians do.

When we had traveled about twelve miles in this oak forest, quite wintery and desolate in appearance with its barren trees and black burned soil, we suddenly entered a beautiful, broad meadow, which could be none other than the valley of the San Saba.

The joy which we experienced at that moment was enhanced still more when we saw near us several white tents, which could only belong to the company we were looking for. Involuntarily we urged our beasts on and soon found ourselves in the midst of the members of the expedition, whose astonishment to find themselves visited by white friends in this wilderness was equal to our joy at having found them and thereby having reached the immediate goal of our trip.

The expedition had halted at this place for the past three days. When they were still several miles from the valley of the San Saba, a deputation of Comanche Indians had met them and inquired the purpose of their coming. Later, on entering the valley itself, a royal reception was accorded them by the Indians. About two hundred (of whom about eighty were warriors) dressed in festive attire had arranged themselves on a hill in military formation. After Herr von Meusebach had ridden toward them and by discharging his rifle had given proof of his confidence in them, mutual greetings were exchanged. After this scene, which according to the unanimous report of those who witnessed it was very picturesque, even imposing, the whole company was escorted to their present camp by the Indians. Here a number of presents were distributed among the chiefs, and all had lived in the best accord up to this time.

The appearance of the camp, lying in a bend of the river, was very colorful and to a European eye interesting enough. To begin with, the three covered wagons which had been drawn into the center of the camp, were an arresting sight in this pathless wilderness, in which up till now no wagon very likely had entered. Around these the tents had been erected and in front of them whites and Indians mingled in a motley crowd. Even the whites were of diverse appearance and of mixed origin. In addition to a number of unaffected Germans with genuine peasant features, one noticed in the immediate vicinity a group of Mexican muleteers with the unmistakable southern facial expression; then there were a number of American surveyors, equally peculiar representatives of a third nationality, which von Meusebach carried with him in order to point out to them the land to be surveyed. Among these various groups of whites the Indians mingled in great numbers, some to offer articles in trade for food, others cowering near the fire and watching with interest the preparation of the food, hoping to receive a share of it.

A young man about eighteen years old dressed in leather, like all Indians, drew our particular attention. Although he acted like an Indian, his features showed plainly that he was of different origin. He had blond hair, blue eyes, and the sharp profile of the Anglo-Saxon race. We learned from our interpreter that he was born of American parentage and that his name was Lyons. About ten years prior to this time, he had been captured by the Comanches in a raid near Austin, in which his parents had been murdered. Since this time he had remained with the Indians. We tried to persuade him to accompany us to the settlements and to visit his brother who was farming near Austin. But he did not entertain our proposition and assured us several times that he liked his present condition very much and that he had no desire to return to the pale faces. Since he could still speak a little English, we learned a little more about the tribe he was living with, although he was very reluctant to give us this information since he had already fully acquired the mistrustful, laconic mode of expression of the Indians. A little Mexican boy about eight years old, who rode behind him on the horse and whom he treated as a slave, aroused my compassion, for he looked half-starved and was shivering in the cold north wind because of his scanty dress. In answer to my question how he had come here, the "Indianized" Anglo-American answered quietly: "I caught him on the Rio Grande." This was said in a tone of voice, as if he were speaking of some animal.

A twelve year old Indian boy with pretty features and friendly manners, aroused my curiosity on account of the fluency with which he spoke English. I found out through Mr. Neighbors, that he was the son of a chief who had perished in San Antonio in a bloody encounter of which I will relate later. He had been kept prisoner in San Antonio a long time after this happening, but was later returned to the Comanches in accordance with the terms of peace between the Texans and the latter. Despite his youth, he was treated with marked respect by his tribe which indicated the high rank his father had held. He had received a white shirt from one of the companions of Herr von Meusebach and had donned it immediately as an outer garment, evidently regarding it as a very valuable ornament.

In the afternoon several friends and I visited the camp or tent village of the Comanches which lay near our own camp, however, on the opposite side of the river. After crossing the San Saba which was about forty feet wide and two feet deep at this point, and similar to the Pedernales and Llano, flowing rapidly and clear, we saw the tents and huts standing about in irregular order. Probably several hundred horses were grazing round about us. The wigwams were serviceably and comfortably made. They are round, twelve to fourteen feet high, and made of tanned buffalo hides sewed together and spread over a framework made of poles sixteen to eighteen feet long, crossing each other on the top. Near the ground is an opening which serves as a door, but which is usually closed by a bear skin. On the top is a small slit, which can be protected against pressure from the wind by two flaps, ingeniously arranged, and which serves as a vent for the smoke. The door and the vent of the tents always faced east, opposite the prevailing winds during this time of the year.

Soon we had an opportunity to enter one of these tents to see the interior, as several of the occupants invited us to do so with friendly nods. Upon entering one we were urged to sit down on buffalo and bear skins, which had been spread in a circle on the ground. Thus we had an opportunity to examine the arrangement of the house leisurely. The master of the hut sat in the rear, opposite the door. At his side were his wives, either engaged in caring for the children whom they fondled lovingly, or doing beadwork. The latter, done on strips of leather, was intended exclusively for the men, for unlike civilized people, the men and not the women among the Indians lay particular stress upon outward finery. The poor squaws are the slaves of the men, and forgetful of themselves, they are concerned only to adorn their lord and master and to gain his approbation. In the middle of the tent was maintained a little fire in a hollow which was adequate to warm the tent and at the same time served to roast the meat. A leather rope fastened to a stake driven into the ground and reaching up to a point where the poles cross each other, was evidently placed there to give the tent more stability and to keep it from being blown down during a strong wind.

We were able to make ourselves understood only through signs and the few Spanish words with which our Indian friends were well acquainted. They produced a piece of paper from a leather container which they asked us to read with a triumphant demeanor, as if it was something very valuable. It was a testimonial from an agent of the United States Government, stating that the bearer had conducted himself peacefully and as a friend of the whites during a meeting above Torrey's Trading Post, which was held for the purpose of consummating a treaty. It also contained the request that all the whites who met the bearer, should treat him as a friend. Such *testimonia morum* were shown us later by others and we always found that the possessors regarded them as something very valuable.

Other huts which we visited later resembled the first one and differed only as to size, according to the rank and wealth of their occupants. The wigwams of the Comanches deserve particular mention since they are so serviceable and comfortable. I am sure they afford more protection in inclement weather than many of the ordinary log houses found in Texas. This shows that the Indians possess an inventive mind when one considers that these tents are so arranged that they can be easily dissembled, loaded upon pack horses and transported.

Far into the night we lingered about several large campfires. In addition to the German songs sung by members of our expedition, the Indians also gave us samples of their chants which were rather monotonous and unmelodious.

Chapter XX
Meeting with the Comanches—A Kickapoo Camp—
A Colony of Prairie Dogs

A council was held with the Indian chiefs on the following morning, February 11. All sat down on skins which had been spread out in a circle in Herr von Meusebach's tent. Jim Shaw, our Delaware chief, whose services as well as those of Mr. Neighbors had been accepted for the duration of the expedition, acted as interpreter. The negotiations began after the peacepipe, from which each one took two or three puffs, had made the rounds twice. The conversations were carried on in short, terse sentences, which were immediately interpreted by Jim Shaw. Herr von Meusebach told the chiefs the following: He had come with his people on the peace path to view the land and to greet them as friends. They would also be received as friends when visiting the cities of his people. He now desired to go up the river to see the old Spanish fort. Upon his return from there he desired to have a council with the principal chiefs, Santa Anna, Buffalo Hump and Mope-tshoko-pe (Old Owl) in order to tell them of his further intentions.

One of the chiefs replied with great dignity as follows: The hearts of his people had been alarmed when they had seen so many strange people, who had not previously announced their coming, and whose intention they did not know. But now, since they were assured that they had come as friends and had declared the purpose of their coming, all was well.

Thereupon a number of presents were laid before the highest ranking chief and he distributed them among the other chiefs and warriors. The chiefs received red and blue woolen blankets, thick copper wire for making bracelets, cotton goods for shirts, and tobacco. The common warriors were given red and blue strips of cloth to serve as a breech-clout commonly worn by all Indians, and some tobacco.

After this meeting we bade adieu to our Indian friends numbering about two hundred, among whom there was no notable chief, and started in the direction of the old Spanish fort. We, however, exacted a promise from them that they report our presence to the highest ranking chiefs and inform them of our desire to have a council with them upon our return.

It was our intention from the very beginning to visit this fort, since there was a persistent rumor among the Texas settlers that the Spaniards had worked some silver mines in the vicinity of the fort. The circumstance that the Spaniards, who were very circumspect in their colonization projects, had also established a permanent place there led us to believe that this place offered peculiar natural advantages. We had no definite knowledge of the location of the fort, for neither Mr. Neighbors nor our interpreter had penetrated as far as this point. We had learned, however, from the Indians that it was several days' journey up the river. We traveled only four miles on the first day and halted when we found good grass for our horses. The appearance of the valley was the same as where we had first entered it. The bottom was about one mile wide at this point. Scattered mesquite trees grew on this plain everywhere and the tender grass with its green sprouts near the ground formed a dense lawn. The fact that even in winter food is to be found here for the horses, made the San Saba valley the favorite camping ground for the Comanche Indians at this time of the year.

On February 12, we traveled ten miles farther along the river over rocky hills and halted in the most beautiful camp we had encountered so far on our journey. It was a small prairie bounded on one side by the river and on the other by rather steep slopes. It was covered with scattered live oaks and patches of low shrubs, which gave it a park-like effect.

We had scarcely set up our tents when an old Indian arrived who announced himself as the messenger of Mope-tshoko-pe, one of the three great chiefs, having been sent by him to secure information concerning us. After Herr von Meusebach had again informed him of our peaceful mission and of our desire to meet the three chiefs, he was satisfied and was dismissed with a few presents.

Much difficulty and delay was experienced on the following day in moving the wagons over the stony, pathless hills. Since it was foreseen that these difficulties would multiply when ascending the valley, it was decided to send the wagons and the greater part of the men back to Fredericksburg and to continue the trip with a small mounted company and a few pack mules. This was deemed all the more advisable since the Indians had shown such a friendly disposition, in view of which even a small company had nothing to fear.

Preparations were therefore made on the next day, February 13, to carry out these two different journeys. This gave me an opportunity to acquaint myself with the geological condition of the immediate neighborhood. Near our camp were horizontally-lying beds of a half crystalline, not very hard, grey limestone exposed on the slopes of the hillside. In spite of the crystalline granular structure, numerous well-preserved fossils were contained in some beds of this limestone. Various new species of trilobites predominated here. A kind of *Orthis* occurred less frequently. The same limestone, belonging to the older or silurian division of the Transition formation was also exposed in several places in the bed of the river.

On the following morning, February 14, the two companies of our former expedition departed in opposite directions; the larger one going in a southwestern direction toward the Llano, the smaller toward the ascending valley of the San Saba. The smaller company, which also I joined, was composed of Herr von Meusebach and his excellent servant Schmitz, who among other duties had charge of the culinary department, and despite great difficulties managed it very successfully. There were, furthermore, five educated young Germans, Mr. Neighbors and his American companion, then Jim Shaw and the other Delaware Indian previously mentioned, two Mexican muleteers who had charge of

packing and driving the eight pack mules, and three Shawnee Indians who had been engaged by Herr von Meusebach
as hunters for the duration of the trip. The expedition consisted of seventeen persons.

Since we could not follow the immediate course of the
stream, we ascended the steep slopes. The land assumes
there the very definite character of a plateau or tableland,
so that when one stands on the heights, all the other distant
heights are on the same level and the incised valleys remain
quite unnoticed. These elevations are everywhere rocky and
sterile, however, a sparse growth of grass is always found.
Here and there a stunted live oak or some other kind of oak
had taken root in the rocky soil.

Several times the repacking of our mules, in which the
Mexican showed great skill, caused considerable delay during our trip on this day. We only covered eight miles. We
selected our camp in a wooded section of a small tributary
valley of the San Saba as a protection against a sudden attack of Indians. It is their custom to arrange these attacks in
a manner that they can suddenly appear on their horses and
strike with lightning-like rapidity, thus spreading confusion
and terror among the attacked.

This same place had at one time served a band of Kickapoo Indians for the same purpose. The framework of their
tents, made of bent twigs like a bower, was proof of this.
Unlike the Comanches, they do not carry their tent poles
with them, but only the deer skin with which they cover the
framework erected at every new camp. Nearly every Indian
tribe in Texas has, moreover, some peculiarity about the
construction of their wigwams which would scarcely be
detected by the European novice, but which furnishes the
Texas colonist reliable identification marks.

The Kickapoo Indians, as well as the Shawnees and Delawares belong to the semi-civilized tribes, who have their
permanent home farther north in the state of Arkansas, but
who visit Texas and particularly the valley of the San Saba
and the Llano. They come in small bands, primarily to hunt.

Of all Indian tribes they are the best hunters and in their beautiful, well-chosen camping grounds one can find heaps of deer hair and bones of bears and other animals. The Comanches look upon these visits with envy and jealousy, but they do not dare to drive them out by force, since they recognize their superiority as marksmen and therefore fear them.

We advanced only twenty miles on February 15, for the rocky terrain over which our road led us, tired our American horses (used to the lowlands and to a generous ration of corn, without which they had to do here) to such an extent that we moved forward very slowly. We learned on this expedition that for such journeys into the mountainous regions of Texas, the little Spanish ponies or still better, mules are far more serviceable than the large, heavy American horses. For no matter how much they are able to do when fed plenty of corn, they soon become exhausted when dependent only upon the prairie grass. The little Spanish ponies and the mules, however, thrive well on grass and require no special care and attention. The contention that, due to their fleetness, one can make his escape more easily on an American-bred horse in case of a sudden Indian attack, does not hold true here, since it would have been impossible to ride swiftly over the rocky terrain.

While crossing a small brook, one of the Shawnee Indians killed a peccary or Mexican hog (*Dicotyles torquatus L.*) As soon as it was killed, the Indian removed the musk gland on its back to prevent it from contaminating the meat. If this precaution is taken, the flesh of these animals is palatable and similar to that of our domestic hogs, as we had occasion to find out that same night.

We camped in a narrow valley bounded by steep slopes and cliffs. This place had the three qualifications for a good camp—water, grass for the horses and wood. Since it had developed during our day's journey that several pack mules were too heavily laden we had to find ways and means to lighten their burden. To accomplish this purpose it was decided to drink part of the wine supply which we carried with us. While carrying out this resolve, sympathy for the poor mules increased until the last bottle of our supply was emptied. The general happy feeling which developed in consequence of this, found expression in the singing of German songs, among which Arndt's excellent patriotic song, "Which is the German's Fatherland," was one of the first to be sung. While sitting around the campfire, blazing high in the sky, these songs resounded in the rocky valley (which had not been trodden by the foot of a German) despite all the Comanches and other wild riff-raff.

Our Delaware chief, Jim Shaw, who had also been given an opportunity to express his sympathy for the mules, felt the urge to give us a sample of an Indian song. He laid down on his back, and while slapping himself rhythmically with the palm of the hand in the region of the stomach, he uttered monotonous sounds which sounded to ears, not trained to appreciate Delaware music, more like the plaintive tones of an individual suffering with a severe case of bellyache, than anything else.

I had slept several hours after this prairie banquet in a little tent which Herr von Meusebach shared with me, when I felt someone shake me. Looking up in the darkness which was dimly lighted by the dying embers of our campfire, I saw the form of an Indian bending over me. This apparition would have frightened me extremely if I had not concluded at the same moment, that this must be a very humane kind of Indian who first wakes his victim before despatching him. Just then he whispered to me, "You, big chief," and I recognized in the shadowy form the servant of Jim Shaw. I answered in the negative and referred him to Herr von Meusebach, sleeping near me. To him he disclosed that his master could not find peace and rest and that he urgently begged for another bottle of firewater. Herr von Meusebach called his majordomo Schmitz, who satisfied the desire of our Delaware friend.

On February 16, we traveled about fourteen miles over rocky hills, and camped on the north bank of the San Saba

River. The barren, sterile character of the heights over which we passed was the same as that which we had previously passed. Also the rocks which formed the terrain were apparently of the same Transition limestone which we had seen farther below. However, this was only a lithologic similarity, for some fossils found in various places showed that we had again come to beds of the cretaceous formations. It is noteworthy that here a much younger series of beds takes the place of an older at the surface without a change of any sort in the natural aspect of the area. A species of cactus (*Cereus Roemeri Muehlenpfordt*) different from any I had seen so far, was quite common in several places. Toward evening we descended from the high plateau into a beautiful prairie in which tender grass and mesquite trees grew. This was the first large plot of good land we had seen since leaving the first camp on the San Saba. Up to this time the valleys were very narrow so that the sterile approaches in most instances extended to the river itself. Here the bottom was perhaps a mile wide and extended several miles upstream.

It was on this prairie where most of us saw for the first time a colony of prairie dogs. A number of mounds, two to three feet high, covering an area of about a half-mile in diameter, were scattered about. The grass round about them was destroyed. On the top of each elevation was a round opening which served as an entrance to the subterranean passages connecting the homes. On each mound sat a little yellowish grey animal, the size of a rabbit, which on closer approach emitted a whistling sound and disappeared quickly in the hole of each mound.

The entire company immediately felt the urge to kill a prairie dog, in order to learn a little more about them. But this proved to be more difficult than it appeared on the face of it, for the animals were so wary that we could hope to shoot them only with a rifle. Furthermore, they usually had time, even when mortally wounded, to withdraw into the depth of their houses to prevent their dead bodies from falling into the hands of the enemy. Finally one of the Shawnee Indians, spurred on by an offer of a reward, managed to kill one of these animals, and we now had an opportunity to satisfy our curiosity.

The prairie dog was about twelve inches long, and resembled our marmot in appearance. The fur was soft and of a yellowish brown color. As is known, this animal belongs to the order of marmots (*Arctomys Ludoviciana*) and the name "prairie dog" is really a misnomer. The thick head and the playful antics, observed from a distance, do remind one of a young dog at play and probably that explains the origin of the name. These animals are plentiful in the northern prairies between the Mississippi and the Rocky Mountains, and all travelers who travel from Missouri through these oceans of grass to the mountain region of New Mexico, describe these colonies as something out of the ordinary, since they break the monotony of this long trip. Their presence in latitude 30° N.L. had not previously been recorded in Texas. They are not found in the lowlands. Their presence in the San Saba valley can be explained on account of the higher elevation and therefore cooler climate. The claim made frequently that prairie dogs hibernate in winter just like the marmot does in Europe, is not true, for Lieut. Abert, in his 1848 report, made mention of meeting them on his trip from Santa Fe in New Mexico to St. Louis in Missouri in the month of January, moving about quite lively in the snow and ice near the entrance of their burrows.

We pitched our camp on the forested bank of the river. The water had a depth of forty feet here, but scarcely a perceptible fall. It is a rather peculiar phenomenon that in all streams and brooks of western Texas, the depth and the fall vary so much. Very often one finds brooks flowing rapidly, but containing a small volume of water in one place, whereas a little farther down they form a basin ten to twelve feet in width, where the water does not seem to move at all. The brooks at New Braunfels and Fredericksburg offer a definite example of this and in like manner does this condition obtain in the larger streams, such as the Guadalupe, the Pedernales and the Llano. The following explanation is about the only one which can be offered for this phenomenon. Considerable floods following a heavy downpour of rain at certain times of the year, deposit great masses of gravel and earthy material in certain places of the stream bed which form natural dams for the water above. These are too considerable for the erosive power of the stream, flowing peacefully again and striving for the attainment of a smooth path of fall, to be removed during the intervals between such floods.

Since deer, turkey and wild geese (*Anser Canadensis Lin.*) were plentiful in the forested banks of the river where we had chosen our camp, and the river contained many large catfish, we decided to rest for a day. February 17 was therefore spent in affluence. In the course of the day, a mounted Comanche arrived, who was a straggler, as he himself admitted, from a larger band just returning from a marauding trip into Mexico. He received an abundance of food which he seemed to desire, and after he had exchanged his leather jacket for an old silk coat with one of the men of our company, he took leave of us, apparently satisfied, and continued his journey downstream, but not before he had donned his newly acquired garment which he wore over his naked body.

In the fertile prairie near our camp grew a species of cactus with long, sharp joints and numerous flattened thorns (*Opuntia exuviata Pr. Salm.*) which we had nowhere observed in the lower regions. This again furnishes proof that not all cacti are confined to sterile, rocky, soil.

Chapter XXI
Arrival at San Saba Mission—Fabled Silver Mines—
Preparing Venison

Our journey led us through several small beautiful prairies on February 18. We had traveled about six miles in this manner and did not suspect that we were so near our destination, when suddenly an object resembling an old wall loomed up in the distance among the mesquite trees. We approached it and suddenly stood before the ruins of a large building; we had reached the old Spanish fort or Mission on the San Saba.

Our first impression was one of astonishment to see in this wilderness, in which we had traveled so many days, far removed from the habitations of civilized man, the indisputable evidence of a former permanent abode of the white man. We entered the inner courtyard through a breach in the wall and found in it a suitable place to pitch our tents. The fort lies close to the river on its left or north bank which here is about twenty feet high. The ruins consist of remnants of walls, five to six, and in some places fifteen to twenty feet in height.

The original plan of this establishment is still readily discernible. The outer walls enclose an area of which the shorter side facing the river, is three hundred feet long, and the longer side three hundred sixty feet. In the inside of the outer wall were a number of rooms or casemates, eighteen feet deep with an entrance leading into the courtyard. There were about fifty such rooms in the circle of the courtyard. The main building contained seven rooms and a courtyard whose walls were still intact up to the crossbeams. It stood in the northwest corner of the establishment. The main entrance of the fort was on the west side, but a smaller exit also led to the river. On three corners of the fort were projecting towers for defense, and in the northwest corner stood a larger round tower. The ashlar stones, of which the walls were composed, were bound together by earth. However, on the walls of the main building one noticed mortar. The general plan of the whole establishment is similar to the missions at San Antonio, however, the churches which at these missions as well as those in California are always the largest and most elaborate, this being in harmony with the purpose of these institutions, viz., the conversion of the Indians, was lacking entirely or was very small and insignificant. Nor does the land in the vicinity of the fort show any sign of having been cultivated and in particular one finds no trace of irrigation ditches which are found at all the other missions. In view of these circumstances, it is very doubtful whether there was a mission here at all. It may have only been primarily a fort to gain a foothold in the San Saba valley. Regarding the final state of this fort, mere legends are known in Texas. According to these the fort was supposed to have been starved into submission in the last quarter of the previous century by the Comanches; the Spanish garrison was put to death and the buildings destroyed. Without doubt, documents are still to be found in Mexico which could give definite information about this fort.

Tall mesquite trees and cacti (*Opuntia frutescens Engl.*) as tall as a man, growing in the courtyard and the casemates attest to the fact that the rooms have not been inhabited for generations. On the portals of the main entrance we found the names of the few visitors who had visited the place during this century. We found the following names: Padilla, 1810; Cos, 1829; Bowie (con su tropa), 1829; Moore, 1840. The first two were Mexicans, the latter Texans who probably had come as far as this region on a military expedition against the Comanches.

On the following day, February 19, we inspected the vicinity of the fort. The site was well-chosen. A level plain with very fertile soil and most tender grass extends on both sides of the river. The width of the bottom is about a mile. It extends for five or six miles along the river and is probably the largest area of suitable land for cultivation in the region of the upper reaches of the San Saba. A strip of woods which could supply the needs of a small settlement is found along the course of the river. No other forests are to be found here and especially is the post oak lacking which furnishes the best material for fences. Farther down we had found extensive forests of such post oaks. At the fort the river is deep, crystal clear and it flows rapidly. However, one can ford it conveniently at several places. It has tributaries here on both sides.

We ascended one hundred fifty foot bluff on the other side of the river which had very steep slopes. From its summit we could see for miles round about us; we were standing on the highest point of the plateau. At no place did we see mountain ranges or individual mountains. There is no San Saba mountain range, although most maps indicate one.

According to legends still told in Texas, the Spaniards were supposed to have worked some rich silver mines here, and the old fort was supposedly erected to protect a mine nearby. Our purpose in coming here was not only to investigate whether the soil is suited for farming, but also to determine if the reports concerning the silver mines were true. We therefore first looked for smelting ovens and heaps of slag in the vicinity of the fort. When we failed to find any traces of either, we made examinations to determine if the geological conditions were of the kind which would indicate the presence of precious metal.

On the rocky slopes of the previously mentioned valley wall, a good section of the rock formation was well exposed. It consisted of alternating beds of rather hard, yellowish grey limestone with rough earthy fractures and soft marl of the same color. All these layers lie completely horizontal, so that for example, as is distinctly to be seen, a firm bed which forms the uppermost member of that section, could have been quarried without difficulty on the level area of the plateau for the building of the fort. These beds contained numerous organic remains. They were of the same species contained in the calcareous marl at Fredericksburg. All were distinct forms of the cretaceous formation and left no doubt with reference to the age of the layers in which they occurred. This same rock also composes the terrain in all directions for many miles around the fort, although it is less distinctly exposed. But with such a relation to the geognostic bedding, one may make the claim without hesitation, that at least in the vicinity of the fort no deposits of precious metals are present. The presence of silver ore in unchanged, horizontal limestone and marl of the cretaceous formation would be without precedent, according to our present-day experience. What has been said is not to be construed as denying the possibility that there may be ore on the San Saba; on the contrary, it is not improbable that the changed Transition rocks, containing quartz, which appear about forty miles downstream near the mouth of the San Saba, as well as the granite rocks which appear particularly between the San Saba and the Llano, contain ore, although observations on our journey did not furnish direct proof for this.

When we returned to our camp, one of the Shawnee Indians had just shot an animal about the size of a marten with a long tail, ringed grey and black. I had not seen it in the lower coastal region, but it does not seem to be uncommon in the higher, hilly country. The Texas settler calls it civet cat, but the scientific name is *Bassaris Astuta Lichtenstein*. It had previously only been reported from Mexico. According to the naturalist who described it, it is a distinct species forming a link between the martens and the "Viverren."

On the following morning, February 20, we made an excursion about six miles above the fort during a very cold norther. The path we followed was well traveled, or it was the war path of the Comanches into Mexico. It crossed the river several times whose volume was quite diminished here. It also led us through several small, pleasant mesquite prairies. The bottom, which was only a mile to a mile and a half wide at the fort, became narrower. The high bluffs on both sides flattened out and one could plainly see that the river valley toward the upper course was on the same level with the plateau. The latter extends westward from the springs of the San Saba (which according to Jim Shaw, were only fifteen miles above the fort) to the Rio Grande or Rio Bravo del Norte. This is the plateau in which all the large rivers of western Texas, the Colorado, the Guadalupe, the Medina and the Nueces with their various tributaries have their source. I was assured of this by surveyors with whom I was acquainted who also remarked that a limestone formation is the predominating rock of the tableland.

On the same day we began our return trip down the river and arrived at a place where our horses found abundant grass. Whereas we had remained on the left or north bank of the river on our journey to this place, we now tried to travel on the right side in order to become acquainted with it also.

We traveled twenty miles over beautiful mesquite prairies on February 21. We again passed a village of prairie dogs, and since we had also seen a large village near the fort, it is evident that they are very plentiful on the San Saba. Exposed places on the slopes of the valley showed us layers of white chalkmarl which contained many fossils.

On February 28, we followed the course of the river for fifteen miles without finding it necessary to climb the steep stony banks. In place of beds of the cretaceous formation, horizontal, hard, grey Transition limestone in thin beds appeared fifteen miles below the fort. The depositional relation to the cretaceous bed was not readily discernible. Near our camp close by the bank of the river, we found several well preserved fossils which left no doubt as to their geological age. On our journey one of the Shawnees shot a cougar (called "panther" by the American settler, and "leon" by the Mexican). Our Mexican muleteers who added many things to the store of edible food which the American, who in this respect is quite fastidious, would despise, roasted some of this meat at the fire and ate it with the best of appetite. It was white like veal and resembled it in taste, but was soft and insipid as is the flesh of all carnivorous animals.

On the stony slopes near our camp grew a bush four to six feet high, with plumeous, leathery evergreen leaves. Our

Indians broke off several of its branches, dried them quickly over a fire and then smoked the crushed leaves like tobacco. Several of our company followed suit and found a mixture of these leaves with tobacco smokeable. The bush was a kind of sumac (*Rhus*), up to this time not described, which is also found frequently on the rim of the stony plateau near New Braunfels. The American Indians moreover are supposed to have two more plants of this species which they put to the same use.

On the advice of our experienced interpreter, Jim Shaw, we remained here also the following day, February 23. We occupied ourselves in drying venison since it was a foregone conclusion that we would find very little food in the camp of the Comanches and we would have little time to procure our daily supply through hunting while the negotiations were in progress.

Our hunters did not succeed in killing a buffalo, although fresh tracks were seen nearby, moreover during the whole expedition we saw only a few of these animals and shot one. The presence of the numerous Indian bands at this time of the year on the San Saba River may have driven them away.

The drying and curing of the deer's flesh was done in the following manner: Four forked sticks, an inch thick, were driven into the ground to form a square, so that the forked ends were about two and one-half feet from the ground. The forks were then fastened with other sticks and thin staves were laid across them. The meat, cut into very thin slices, was spread out upon the staves and a small glowing fire was maintained under it. The warmth and smoke dried the meat in one day to such an extent that it could be preserved for some time. After we had prepared the flesh of four deer in this manner, our existence was assured for a few days, although this dried meat was a poor substitute for fresh-killed game.

Our march of twenty-three miles on February 24 led us over rocky, arid heights. It was impossible to follow the course of the stream, since the steep rocky banks reached to the river's edge. On this plateau we saw several times a large, bluish grey species of rabbit which was unknown in the level regions of Texas.

Toward evening we came to a tributary valley of the San Saba, covered with oak and other trees, which presented a pleasant change when compared to the barren heights we had traversed during the day. We camped at the same place where we had parted from the wagons and a greater part of the original company.

After we had remained here several days which was very necessary, due to the increasing weakness and exhaustion of our horses, we continued our journey downstream on February 26. The elevations over which we had to ride for fifteen miles, were just as unproductive and monotonous as those we had seen formerly and we were glad when we reached our camping ground which was again pitched near the tents of the Kickapoo Indians. Near our camp were beds of Transition limestone steeply uptilted and strongly altered. These uplifts and alterations, no doubt, have some connection with the presence of granite rocks which a creek near our camp, filled with decomposed granite, indicated.

Two of our Shawnee Indians left us here for, as they remarked, they did not wish to meet the Comanches, to whom their many deer hides would be offensive. They had hired themselves to us as hunters for a dollar a day with the proviso, that they could retain the skins of the animals killed. In the course of our journey they had accumulated a considerable number of deer hides. The skinning was done with great dexterity and dispatch, but they invariably left a thin layer of muscle remain on the skin. When asked why they did this, they answered naively that the weight of the skin, according to which it was sold, was increased by this. The fact moreover that the Shawnees shot the deer primarily on account of their skins proved disadvantageous for us, for when meeting a herd, they always killed the stags, since they had larger and heavier skin, and not the does, whose flesh was much more tender.

At this point the valley of the San Saba became only a cleft without any river bottom and remained as such for a stretch of twenty miles. In consequence of this on February 27, our journey led us entirely on the plateau, which was more desolate, sterile and stony than any place we had seen on our entire journey. This same monotony was in evidence as far as the eye could see. The distant hills appeared white, as if covered with a light covering of snow. The entire surface was thickly covered everywhere with fragments of a firm, white Transition limestone which formed the foundation here. The whole aspect of this Transition limestone, particularly the frequent appearance of druses of smaller quartz crystal found in them, reminded me very much of the lead ore bearing silurian limestone which I had learned to know on the Merrimac in the State of Missouri a year and a half prior to this. After traveling several hours over these desolate heights, a dim Indian trail which we followed led us toward the perpendicular bank of the San Saba. About a hundred feet below us we saw the river flow turbulently in a cleft, and it appeared an impossible feat to cross it. But the sharp eye of the Indians had seen, that the terraces of the individual layers of stone formed natural rock steps, on which one could descend to the river. By leading our horses, we descended without trouble and after crossing the rapidly flowing, but not very deep stream, we ascended the other side on similar steps. The walls of this cleft in which the San Saba flows, was covered with an evergreen growth which was very pleasing to the eye. Cedar bushes (*Juniperus Virginiana L.*) predominated, but in addition to these there was "Lignum Vitae" (*Dermatophyllum speciosum Scheele*) with its beautiful flowers and an evergreen sumac previously mentioned. Also the shrub of the common *Opuntia* with leaves as large as a hand grew luxuriantly out of every cleft in the rocks.

Over less sterile heights on the left bank of the river we finally reached our camp, which was again an old abandoned Kickapoo camp. These Indian camps appeared at regular stations, usually in pairs a day's journey apart, along the San

Saba, and were, as remarked previously, chosen with great care and in the most suitable and beautiful places. Here in a tributary valley of the San Saba, on its left bank, a succession of beds were exposed on a rock wall, which we had not observed on our journey. They were beds of black, hard limestone, containing large elliptical masses of black chert which can be distinctly designated as carbonaceous limestone through the numerous organic remains.

That the presence of this carbonaceous limestone determines with a degree of certainty that there is also coal present, should be mentioned here, for this may prove very important should this region be settled in the future, owing to the scarcity of wood. One of our company shot a female opossum here. In its pouch were found six immature, naked young, scarcely three-fourths inch long, which lived for several hours after the death of the mother, firmly fastened to its teats.

CHAPTER XXII
A BEE TREE—TREATING WITH THE COMANCHES—
INDIAN PUNISHMENT FOR INFIDELITY

We had scarcely gone a short distance on the following day, February 28, when the cry: "A bee tree! A bee tree!" brought us to a halt. Mr. Neighbors, who possessed the keen sight of the pioneer, had observed the flight of some bees to an old live oak tree. Upon approaching the tree, he had discovered the hole which served as an entrance to their hive. Our entire company halted, for no real backwoodsman will pass up a bee tree without robbing it, no matter how much he is in a hurry. The opening which happened to be on a horizontal thick branch of the tree was soon enlarged with our axes. Axe blows followed in quick succession and the chips flew. Suddenly a shout of joy broke forth, for some of the golden honey began dripping upon one of the men wielding the axe. After a few more strokes, rows of honeycomb filled with honey became visible. Everyone ate as much of the sweet honey as he desired, but despite this there remained enough to fill a whole bucket. Several combs were left in the hive to serve as food for the poor bees in winter, a custom followed by the humane settlers.

The opinion is prevalent in the western part of the United States, that the bees are not original inhabitants of the wilderness, but that they advance with the white man as the buffalo retreats before him. I had occasion to verify the latter statement relative to the buffalo in Texas, but I could not convince myself of the truth of the former. In the place mentioned above and in a number of other places, we found bees in the middle of the wilderness many days' journey from the settlements. As a matter of fact it is hard to understand that the extension of their territory should be dependent upon the presence of the white man.

We had traveled about ten miles along the ridge, when an Indian path led us into the valley. We concluded from the numerous, fresh horse tracks that an Indian camp was near. We soon found evidence of a recently abandoned camp in a pretty prairie along the river, for the grass was either trodden down or cropped short. A few tent poles, left behind, also indicated this. Two mounted men appeared soon after in the distance, who rode away hurriedly in the opposite direction upon seeing us. We followed them and saw, in the bend of the river, a long row of white tents which resembled the military tents of a European army from a distance. While still a half mile distant from the camp, a herald, dressed in his best, and carrying a flag, rode toward us and invited us to follow him. Upon arriving there, we selected our own camping ground at the lower end of the village.

We were still engaged in erecting our tents, when three chiefs, Mope-tshoko-pe (Old Owl), Buffalo Hump, and Santa Anna came to welcome us. We depended particularly upon the latter for a peaceful and advantageous termination of our negotiations since he was favorably inclined to the whites. He had just recently returned from a visit to Washington, where the strength of the white people had much impressed him. The government of the United States pursues the policy of inviting the Indian chiefs to Washington for just this purpose.

In the meantime men, women and children had gathered around us in great numbers and stared at us. They became annoying immediately by pressing around us greedily while we ate and by stealing small articles. This became worse in the following days. The chiefs had given us the promise that our horses would be safe from theft, so we turned them loose. We found them again on the day of our departure which was remarkable evidence of the trustworthiness of the Comanches when they have given their promise of hospitality. This is really remarkable when one considers that such horses as ours were a treasure in the eyes of an Indian, to obtain which he is willing to risk his life. Our Indian friends left us late in the evening, but we were too excited to sleep immediately, owing to the many impressions we had received during the day.

When we awoke on the following morning, March 1, our friends, the chiefs, were already squatting before our fire which they had rekindled, evidently awaiting our appearance. We soon convinced ourselves, however, that they were not so much concerned in wishing us good morning in the Comanche land, but they had a more substantial motive, namely, to partake of our breakfast. We discovered throughout the course of our visit, that their hospitality was more of a negative kind, consisting mainly in the assurance that we would be protected against murder and robbery during our stay, occasional thefts of small articles of course excepted. It was laughable to see how the big chief Santa Anna, a powerful man in the prime of his life, sought to ingratiate himself with our commissary Schmitz by signs and words, in order to obtain a few sweets. In extenuation of our red friends' obtrusiveness be it said that obviously there was a scarcity of food in the entire camp, there being nothing on hand but a little buffalo meat, a circumstance, which is however, not surprising in view of the indolence and carefree nature of the Indians.

We visited the camp or tent village after breakfast. It was much larger than the one we had seen previously. About one hundred fifty tents of various sizes were scattered about in irregular order along the seam of the woods. Several of them, belonging to the principal chiefs or used for public meetings, distinguished themselves from the others by their size and stateliness. Near several of the wigwams were the war emblems of individual warriors, composed of shields, a peculiar headdress made of buffalo skin with the horns of the buffalo fastened to it, and a lance. These weapons were placed on a platform made of stakes and are medicine, i.e., holy, pertaining to their religious ceremonies, and no one dare touch them. As soon as we neared a wigwam, we were received by the surly barking of a number of ugly, lean dogs, which sneaked away cowardly when we walked toward them. The squaws were busily engaged everywhere with their house duties. Some of them were making ropes out of horse hair, used for tying the horses; others braided leather ropes or lassos out of narrow strips of horse skin; still others were preparing the hard buffalo skins for use, using a hook-like short handled instrument to scrape off the fleshy, fat parts of the skin; while still others cleaned skins which had become dirty by rubbing them with white clay until they were white.

A little farther on we saw a squaw with a pack horse laden with venison. She was just returning home and unloading the meat in front of the wigwam. The men kill the game and then send the squaws out to bring the meat home. In another place we saw a number of women engaged in taking down the tents and packing them on mules. Such a mule presents a peculiar sight, with a thick bundle of tent poles, twelve feet long, fastened to its side, dragging along the ground and skins packed on its back. The tracks made by the tent poles dragging the ground are always a sure sign indicating the direction the Indians have traveled.

While we were inspecting the village, articles were offered us in trade on all sides, particularly skins. A good buffalo robe could be exchanged for a woolen horse blanket. Small pelts, for example those of the grey fox and the ringtailed animal previously mentioned, we obtained for a small portion of salt or corn. We also traded a leather lasso for a small quantity of cinnabar. As a general rule, the Indians preferred useful articles or edible things to trinkets or trifles.

About one thousand horses were grazing near the village. As long as they find food and their masters game, the Indians will remain in one locality. For this reason they wander about from place to place.

Most of the horses, as a rule, are unsightly and small. However, we saw some very fine mules which evidently had been stolen on their raiding expeditions into Mexico. The ears of the horses of the Comanches were slit to distinguish them immediately from horses of other tribes.

The council agreed upon with the chiefs took place at noon. A number of buffalo hides were spread out in a circle in front of our tents. The chiefs and the most renowned warriors sat down on one side, while opposite them sat Herr von Meusebach, our interpreter, Jim Shaw, Mr. Neighbors and several others of our company. The three chiefs, who were at the head of all the bands of the Comanches roaming the frontiers of the settlements in Texas, looked very dignified and grave. They differed much in appearance. Mopetshoko-pe (Old Owl) the political chief, was a small old man who, in his dirty cotton jacket, looked undistinguished and only his diplomatic crafty face marked him. The war chief, Santa Anna, presented an altogether different appearance. He was a powerfully built man with a benevolent and lively countenance. The third, Buffalo Hump, was the genuine, unadulterated picture of a North American Indian. Unlike the majority of his tribe, he scorned all European dress. The upper part of his body was naked. A buffalo hide was wound around his hips. Yellow copper rings decorated his arms and a string of beads his neck. With his long, straight black hair hanging down, he sat there with the earnest (to the European almost apathetic) expression of countenance of the North American savage. He drew special attention to himself because in previous years he had distinguished himself for daring and bravery in many engagements with the Texans.

As soon as the negotiations began, the women and children, who had surrounded us up to this time, withdrew to a respectful distance and formed a gay decoration throughout the entire conference. In the center of the circle lay a small pile of tobacco and a pipe. An Indian took the latter, filled it with tobacco and after lighting it and taking two puffs, he passed it around. Twice the peace pipe was passed around without the silence having been broken. Finally Herr von Meusebach made the following proposals through the interpreter: The Comanches should permit the Germans to establish a settlement on the Llano and give permission to survey the land lying north of it, particularly also the valley of the San Saba. For these concessions the Comanches should receive presents to the value of one thousand Spanish dollars two months hence in Fredericksburg where a meeting was to be held. The Comanches were also to be treated as friends whenever they visited the German settlements.

The chiefs consulted with each other in a low voice after the speech was ended. Thereupon, Mopetshoko-pe replied that they were obliged to consider these proposals a little longer and that an answer would be given in the morning. Thus ended this meeting.

In the twilight of the same day we witnessed an odd spectacle, for a number of mounted men appeared before our camp in festive, but most peculiar attire. Their faces were painted red and the majority of them wore the peculiar headdress made of buffalo skin with the horns of buffalo attached. We had seen this displayed before their tents previously. In one hand they carried the long lance, painted red, in the other a round shield made of tanned buffalo hide, painted in gaudy colors and decorated with a circle of feathers which fluttered in the breeze whenever the shields were waved to and fro.

Even the horses shared in the grotesque appearance of the riders, for while they were mostly light in color, their heads and tails were painted a carmine red. The troop paraded several times before us in a slow gallop and finally disappeared in the darkness.

This was a war party of young warriors who were preparing for a raiding expedition into Mexico and who wanted to accord us this attention before departing. The insecurity and the distress in the border provinces of Mexico, particularly Coahuila, Chihuahua and Tamaulipas, where these Indians make their forays regularly, must be boundless. If a strong government will not soon replace the present one in Mexico, these provinces, which under Spanish rule grew and prospered because that government knew how to keep the marauding Indian tribes in check, will be desolate and depopulated, and the Indians, encouraged by their success, will extend their raids into the heart of Mexico. No doubt these provinces will not receive a full measure of protection until (as was the case with Texas, New Mexico and upper California) they have come under the protection of the United States through peaceful or forceful conquest. We saw among the Comanches all kinds of movable articles stolen in Mexico, such as costly blankets, mules, horses, harness, etc. We also saw captured Mexican men, women and children. Several of these had lived for so long a time among the Indians that they had no desire to return to their native land; they were therefore not treated as prisoners. A young Mexican woman with whom her master was dissatisfied, was offered to us for the small sum of forty dollars.

Early on the following morning an old man appeared before our camp and complained to the assembled chiefs with a doleful countenance, that these same young warriors who had paraded before our tent in their warlike accoutrements, had stolen his wife and two of his best horses. The chiefs, to whom the misfortune of the old man seemed to appear ludicrous, advised him to pursue the young men and to recapture his stolen property. The old man returned in the evening with a satisfied look and related that he had found the young warriors not far distant from the camp and surprised them, just as they were in the act of drying the flesh of both of his horses. He had taken two of their mules to replace his horses and had also regained his wife. In fact, the latter accompanied him and was a rather good looking young woman, who had probably become tired of the old fellow. In answer to the query of our interpreter why he had not cut off her nose, he replied that he was glad to have her back and did not contemplate doing any such thing. As punishment for infidelity, the Comanches have a custom to mutilate a guilty woman thus and then to cast her out. We actually saw several such women who had their noses cut off and had short cropped, bristly hair covered with vermin, which they were just picking from one another. Clad in filthy skins and with their long, wrinkled breasts, they presented the most horrible picture of femininity I have ever seen.

The second meeting with the chiefs took place at noon. The negotiations proceeded in a manner described previously. After discussing the matter thoroughly which is characteristic of the mistrustful and cautious nature of the Indians, the proposals made the three chiefs by Herr von Meusebach on the previous day were accepted. The council ended by mutually embracing each other, whereby the Comanches tried to show the degree of their friendship by the strength of their embrace. They were then served a meal of venison and rice which Herr von Meusebach had prepared for them.

Our departure was scheduled for the following day in view of which we received a peculiar serenade in the middle of the night. We were awakened by the confused noise of voices and instruments. When we stepped in front of our tents, we recognized in the darkness, dimly lighted by our campfire a crowd of men and women who began to intone a wild, monotonous song, to which others beat time with sticks on a tightly stretched buffalo skin. Although we appreciated their good intention, this seemed more like a charivari to us than a serenade.

We occupied ourselves on the following morning, March 3, with packing. After having purchased two mules which we needed to carry our equipment, due to the weakened condition of several of our horses, we departed at noon. The majority of us looked forward to this moment with longing; for no matter how great had been the charm and interest to see the Indians, so little known to us, in the midst of their hunting grounds, and to observe their customs and their unadulterated primitive mode of living, nevertheless our stay had be come irksome to us, due to the voracity and obtrusiveness of these sons of nature, as well as their continual thievery. Every one of us missed some article upon our departure. Although I had kept a watchful eye upon my property not less than four articles were missing from my saddle, bridle and clothing. The stealing of small articles was very likely done by the women, for the warriors deemed it beneath their dignity.

According to our original plan, we were to go to the mouth of the Colorado immediately after the council with the Indians; then to follow it in its course to the mouth of the Llano; and from that point to return to Fredericksburg after we had explored the valley of the latter stream. Diverse reasons compelled us not to extend our journey, and to return to Fredericksburg with all possible speed.

According to the calculations of our interpreter, we were about thirty miles from the mouth of the San Saba. The entire course of the river from its source to its mouth was therefore one hundred thirty to one hundred forty miles. We also knew that we could not follow the immediate course of the river, since, as mentioned previously, the steep, rocky banks arose from the bed of the stream at some places.

Chapter XXIII
Description of the Comanches—
Geological Observations—Return to Fredericksburg

Before taking the reader back to Fredericksburg, a few additional remarks about the Comanche Indians may be in place here.

The Comanches, also spelled Cumanches or Camanches, are without doubt the most powerful and dangerous among the North American Indian tribes. Their migrations and marauding expeditions extend over a huge territory. They call themselves the lords of the prairie. Their abode is preferably the rocky tableland between the upper course of the Red River and the Rio Grande. The San Saba valley is preferred as their winter quarters. They have always regarded the Llano River as the southern boundary of their hunting grounds. However, they drift occasionally also as far north as the Arkansas River, according to Gregg's *Commerce of the Prairies*, and fifty years ago are supposed to have lived there. They also wander about in the uninhabited wild region between the Nueces and the lower course of the Rio Grande, and even go down to the coast. Just this year the news has reached Europe of a bloody raid on a Mexican settlement on the Rio Grande near Camargo, where hundreds of people were either murdered brutally or led into captivity. The Apaches, who in conjunction with the Navajoes harass the northwestern provinces of Mexico in a similar manner, particularly Sonora and Durango, are related to the Comanches. They are divided into numerous bands under different chiefs. Their number is estimated to be about 10,000 which, of course, is only approximate.

The Comanches are primarily hunters without a permanent home and carrying on no agriculture. They wander about continually, usually following the march of the buffalo on which they are principally dependent. Year in and year out they eat buffalo meat. It was odd to see mothers feeding their two-year old children dried buffalo meat. The only vegetable food of which they partake occasionally, seems to be the thick node of the root of a plant belonging to the species *Psoralea*. At any rate I saw an old squaw dig up and gather such nodes along the banks of the San Saba in a wooded bottom. It is evident that with such a mode of living, dependent upon the hunt ,and considering the natural indolence of the North American Indian, there is quite often a dearth of food and even want. In such an event horses and mules are slaughtered, which happens quite often when they are on the warpath and have no time to hunt. The fact that they are dependent upon game for their sustenance, prevents a great number from gathering in one place, and herein consists the chief protection of the whites against the Indians, for if all the different bands would unite in a common cause, this would prove to be a terrible calamity for the sparse population of Texas.

They are not only essentially hunters but also equestrians. In all their major undertakings they use horses. They fight, they hunt, they travel on horseback. Their dexterity in riding is extraordinary. This is especially praised by such who had opportunity to observe them in their engagements with the whites, when they would, for example, ride toward their adversary with lightning rapidity, hanging on the opposite side of the horse so that no part of the body was exposed.

The women also sit astride the horses like the men, and ride almost equally as gracefully. The little Spanish horses are used, and although they are not attractive, they have great power of endurance. The Comanches can travel sixty miles with them in one day in the mountainous, stony regions, whereas other travelers would consider thirty miles a very good day's journey. They raise most of their horses themselves, but also capture many on their expeditions into Mexico or steal them from the Texas settlers. They justify the theft of horses in a manner which shows their ignorance. They say it is evidently a great injustice of the "Great Spirit," that he has given the white people, who are in the minority, so many horses, whereas they had received so few. Now they were obliged to balance this disparity as well as possible.

It strikes one as peculiar, when reflecting upon it, that the Indians, whose present condition carries the stamp of originality, led an altogether different life several hundred years ago. For all the peculiarities and conditions of their present mode of existence dates back to the time when they received horses through the Spaniards. Giving them this domestic animal was an important event through which the arrival of the European in America transformed the entire mode of living of these tribes.

The weapons of the Comanches are still the bow and arrow and a long lance. The bow, about four feet long is made of the wood of a tree native to eastern Texas and Arkansas (bois d'arc; *Maclurea Aurantiaca Nuttall*). The light arrows, a little more than two feet long, are carried on the back in a quiver made of horse hide or occasionally out of the skin of cougars. The arrowheads are often made of iron, formerly they were of flint. One can frequently find places in the hills where these were made, indicated by countless fragments of flint or incomplete or defective arrow heads. When examining these apparently crude weapons, one would think they were not very effective and yet the Comanches kill the buffalo with these bows and arrows with the greatest safety, whereas a bullet will not always penetrate every part of the shaggy skin of the animal. At times the arrow is even shot with such force, that it emerges on the opposite side of the buffalo. The long feathered lances have a long iron point, which often consists of an old Toledo sword blade, several hundred years old.

In addition to this we found many Comanches also armed with the long, American rifles, but they did not depend much upon them, neither were they very proficient in their use.

The dress is not very different from that of other North American tribes. It consists of the common leggins, moccasins, the breech-clout or flap, and a buffalo robe or woolen blanket, which covers the body like a mantle. They also frequently wear a close-fitting jacket or shirt, closed in front, made of deerskin. The women are dressed in a short skirt or a kind of tunic made of deerskin, which is often decorated with beadwork or small pieces of metal dangling from it. In addition to this they wear short leggins and moccasins. The hair is cut rather short, the long decorated braid hanging down the backs being the prerogative and pride of the men. They also have no covering for the head, just like other Indian tribes. Although the Comanches prefer deer and buffalo hides as articles of dress, nevertheless many possess, as was mentioned previously, woolen blankets and cotton shirts and other articles of American manufacture which they have received as presents from the government of the United States or obtained in trade at the trading post. In general their dress is less tidy and clean than that of their kin, the Lepans, where the dresses of the women particularly are often quite dainty. But they also love ornaments of beadwork and prefer the blue glass crystals.

As far as their build is concerned, the Comanches are supposed to be better formed and have a more noble expression of countenance than other tribes. I have not found this to be the case from personal observations. The men are usually powerfully built, but not handsome. One seldom finds a well-formed face with regular features. They are inferior in this respect to the half-civilized Delawares and Shawnees, among whom tall, handsome figures and noble features are not uncommon. The women are usually small and thick set, and only in early youth some of them are well-formed and good-looking. They fade quickly, due probably to the hard, physical work which they are compelled to perform, and in old age they often present a horrid picture, as described previously. However, the little children with their coal-black, fiery eyes, the glossy black hair and the brown skin, through which the glowing red cheeks shine, are very pretty, and it is gratifying to see with what tenderness they are treated by their parents. While still very young, they are carried on the back of the mothers, strapped in a peculiar bag made of leather and thin boards, from which the head only protrudes.

The Comanches distinguish themselves from other Indian tribes in scorning the use of alcoholic drinks. As is well known, all other North American Indians are passionately addicted to the use of whiskey, and this hellish drink is carried to them in the form of alcohol by unscrupulous traders, since it is easily transported in this manner. Next to the pocks which destroy whole tribes such as the Mandans on the upper Missouri, alcohol is the chief curse which their acquaintance with the whites has inflicted upon them and which will hasten the inevitable day when it will be said: The Red Race of the North American continent is extinct.

The Comanches not only disdain the use of whiskey but also despise the individual who becomes intoxicated. One time I saw several Comanches watch a drunken Delaware staggering on the streets in San Antonio, and I shall never forget the disgust registered in their faces. Yet how long will it be until they too will succumb to this vice, due to their frequent contact with the whites? Not very long, if the general axiom holds true that the Indians acquire only the bad habits and vices of the whites.

The Comanches are also superior to other tribes in their valor. Whereas others attack the enemy usually from an ambush, the Comanches are not afraid to meet the whites in the open. Many examples of this were furnished in their wars with the Texans lasting over a period of years.

An episode, retold in various versions, but which nevertheless does not reflect credit upon the character of the Texans involved, gives an illustration of their undaunted determination when called upon to defend their freedom. When the Texans under President Lamar's administration had lived in warfare with the Comanches for some time, it was decided to make peace, since the war was beginning to be irksome and no decisive advantages had been gained. The Comanche chiefs were invited to come to San Antonio to conduct peace negotiations and to bring their prisoners along in order to negotiate also for their release. In compliance with this summons, about fifteen chiefs appeared at the appointed time in San Antonio, but the prisoners were left in a camp, several miles distant from the city. The negotiations began and on the first day the amount of ransom money for the prisoners was agreed upon. On the following day the prisoners were not returned, as was promised, but the chiefs demanded a higher ransom than agreed upon.

Wrought up over this breach of promise, the presiding Texas officers declared that the chiefs would be kept as hostages until the prisoners of war were brought to San Antonio. This was contrary to all rules of international law pertaining to the inviolableness of peace negotiators. In the same moment when they heard that they were prisoners, the highest ranking chief sounded the war whoop and at the same time shot one of the Texas negotiators with an arrow through the breast; the rest followed his example and before the Texans could make use of their weapons, several more were killed and wounded. But they were in the majority and an armed force was also stationed outside. These stormed into the building and killed all the Indians except one, although they defended themselves with the greatest bravery. The latter managed to escape and found refuge in a stone building where he defended himself a long time, finally managing to break through the mob surrounding the house and escaping the second time.

When the battle began, the twelve-year-old son of a chief was playing in front of the door of the building. Suddenly he heard the war whoop of his people and immediately he shot down with an arrow one of the Texans who was hurrying to the building.

This story was told to me by an old Texan who justified this act, since it was actuated by the wish to free his captive compatriots from the hands of the barbarians. I myself saw the holes in the woodwork (caused by the arrows and bullets) in the courtroom of the courthouse at San Antonio, the scene of this bloody drama.

Another story is told, interesting also in other respects, which shows what noble sacrifices the Indians are capable of bringing in time of danger. In the year 1841 a band of Comanches appeared suddenly before the little hamlet Linville on Lavaca Bay. The inhabitants, not strong enough to defend themselves, withdrew and abandoned their homes and stores. The Indians seized the rich booty, loaded it on their pack horses and retreated toward the hills as quickly as possible. However, the news of this bold attack spread throughout Texas and in a short time a number of armed settlers, at that time always prepared for such emergency, gathered in order to pursue them and retake the plunder. The Indians were overtaken at Plum Creek near Bastrop, and the battle began at once. Many of the Indians were killed and the booty, composed of cloth, cotton and silk goods, was scattered in the prairie. The rest of the Indians sought to reach the hills in wild flight. It happened during this flight that a squaw who was pursued by several Texans, fell with her horse and was in danger of being captured, although she defended herself bravely. As soon as her husband, who was in the lead, saw the danger in which she was, he returned voluntarily and was killed near her, since he was not able to effect her rescue.

There is a legend among the Comanches, told by William Bollaert in his report to the Royal Geographical Society of London, which sheds some light upon the manner in which they are supposed to have come to this country and which also explains the derivation of the word Texas. According to this legend they are direct descendents of the subjects of Montezuma II and migrated north when Cortez destroyed the old Mexican empire, since they did not want to bow under the yoke of the foreign conqueror. After traveling many weeks they came to a large river (the Rio Grande). They climbed a steep hill on the other side of the river, and when they saw the level land spread out before them, covered with countless herds of buffalo, deer and antelopes, they cried out involuntarily: "Tehas! Tehas! Tehas!" and they decided

to make this country their new home. The word Tehas in the
Comanche language signifies the happy hunting grounds,
i.e., the abode of the departed spirits. Later the Spaniards
changed Tehas to Texas.

So much about the Comanches. However, I would like
to mention that I also saw bands of "Apaches Mescaleros,"
during my stay in Texas, who had their abode farther west
on the San Saba and Llano, but who were considered strangers by the Comanches although they were related to them.

When we started on our return journey to Fredericksburg,
we were at a loss which direction to go, but our interpreter,
Jim Shaw, although he had never been over this route before,
was at no time in doubt, and as was proven, he was right.
This appeared rather uncanny to the European, although it
stands to reason that the many years of training of this sense
of direction has developed it to a remarkable degree.

We left this beautiful valley of the San Saba, and our journey led us immediately for fifteen miles over the most arid,
rocky and desolate region we had encountered so far. Only
after reaching the divide toward the Llano, did the appearance of the country change. We saw a chain of well-formed
round hills toward the south, which with their sharp outlines reminded us of the "Siebengebirge" on the Rhine.

The character of the beds at the surface also changed at
the same time. Instead of the hard, light grey Transition
limestone with many druses of small quartz crystals, and
occasionally fossils (particularly *Euomphalus sancti sabae*),
there came in a reddish Transition limestone, to the eye apparently changed metamorphically, as we had recognized it
previously while crossing the Llano.

We camped for the night in a tributary valley of the Llano
at a waterhole. We rejoiced at the prospect of warming our
limbs, grown stiff from riding a long distance against a
cold north wind, at a huge campfire. Thanks to the Indians,
our evening meal was very meager, consisting of coffee, of
course without sugar and milk, and a little cooked rice.

On March 4 we rode about twenty miles through a region

grown up with post oaks, which was rather low and level, although individual hills and rock masses of red granite arose.

This rock outcropped immediately after leaving our camp and remained the predominant rock on the following days until we reached the divide between the Llano and the Pedernales, approximately twenty miles distant from Fredericksburg. The more level plains between these granite elevations contained a light, red soil (which arose from the weathering of the granite) and in many places a bare, coarse, disintegrated granite gravel. Although the soil was light, it was by no means unfruitful, as was proven by the mesquite trees growing particularly along the course of the creeks. This region must be beautiful in summer when grass and trees are green and would also yield good crops if the season is not too dry. Moreover, the region here is altogether similar to that which we crossed between the Llano and the San Saba on our first trip.

We had scarcely selected our camping ground in the afternoon, when the chief Santa Anna, with his wives and several other Indians, rode toward us with friendly greetings. They desired to accompany us to Fredericksburg and would have gone with us immediately upon leaving the San Saba, if some preparations had not prevented them. Among them was a chief, named O-zanarzco, whose relative had been killed during the bloody episode in San Antonio de Bexar. In the meantime he had avenged himself by murdering several whites near Austin. Until now he could not be persuaded to meet peacefully with the whites or to negotiate with them. Since for very good reasons the Indians could not deplete our food supply, their company did not become vexatious this time. We had hoped in vain to kill a deer during the day as our meat supply was totally exhausted.

When the Indians saw that our days of affluence were over, they themselves went hunting. In a short time one of them returned with a large piece of meat hanging from his saddle. We greeted the prospects for a good supper joyfully, but the majority lost their appetite when they observed a

piece of horse hide hanging from the meat. The Indian had
shot a mustang or wild horse not far from our camp and had
brought a piece of the meat along with him. The flesh tasted
very good and, when roasted on a stick, resembled beef. This
of course can be expected since the mustang, like the game
animals, subsists exclusively upon grass. Even the flesh of
the domestic horse is very palatable.

On the following day (March 5) we arrived at the Llano
after traveling fifteen miles. We remained here since we did
not know if we could find water in another camping place.
We caught a catfish four feet long in the river and as someone also bagged a deer during the day, our food shortage was
at an end. The river was very deep here and flowed rapidly
over its bed of granite rock. The banks of the river would not
be suited for a settlement since the strip of land is too narrow and wood too scarce.

On March 6, we rode twenty miles to another camp beyond Sandy Creek. This creek is the only water worth men-
tioning between the Llano and Pedernales. Its bed is also
cut into the granite and partially filled with granite gravel or
sand from which circumstance it derived its name. Its water was dried up with the exception of a small strip. It is not
separated from the Llano by a higher divide, but between
both lies a low region which, as mentioned previously, was
mostly composed of a light, sandy soil partially covered with
oak trees and on which single granite hills and boulders
arise.

The following day, March 7, was to terminate our journey, according to the reckoning of our Indians. However,
we still had a strenuous ride of thirty-five miles before us
which was very tiresome for me, since I had a recurrence of
fever during the past days, due to the cold northers. At first
our journey led us over low land similar to the one we had
just seen. But soon we saw the divide toward the Pedernales
arise before us in the form of a ridge, probably a thousand
feet high. Simultaneously the granite ceased to be the predominating rock and a white, hard, horizontally lying creta-

ceous limestone came in again in its place. A narrow Indian path wound along the dense undergrowth to the top of the hill along which the squaws could hardly pass with their pack mules. The summit was covered with a continuous oak forest which extended beyond Fredericksburg.

We arrived in Fredericksburg in the afternoon to the great surprise of our friends who had been very uneasy about us and who had even given us up for lost since our absence extended greatly beyond the time of our expected return. They had also drawn an unfavorable conclusion for our safety from the sudden flight of several Comanches who had visited Fredericksburg.

Thus this expedition ended happily. Its result will be of value as far as the geographical and historical knowledge of the northwestern part of Texas is concerned. At the same time it was the first authentic news which the Mainzer Verein had received of the grant designated for an extensive colonization by the Germans.

To Herr von Meusebach goes the credit for having recognized the importance of this undertaking and for carrying it out with determination and great circumspection. With reference to the possibility of establishing settlements on the grant, it must be emphasized that the areas of good fertile soil suitable for agriculture are not very great north of the Llano and on the banks of the San Saba. It does not in any way measure up to the idea generally accepted of the unsurpassed fertility of the region of the San Saba valley. The circumstance that we traveled in winter when all the grass was burned off everywhere enabled us to observe the composition of the soil much better than in summer, although we did not receive as favorable an impression of the beauty of the landscape. The bottom only is suited for agriculture since the steep, rocky slopes in many places rise immediately from the bed of the river. The slopes of the valley and the higher land is everywhere rocky and sterile and the scarcity of water would in itself preclude a settlement here. Several places on the San Saba, particularly in the vicinity of the old fort, are attractive enough to invite the settlers to establish a few homes. However, we did not see an area extensive enough for a large settlement or the founding of a city like New Braunfels. There are also not enough forests here for such a project. It would, of course, be quite an advantage if settlements were established along the San Saba. However, it would involve many difficulties to drive out the Comanches when one considers that they have emphatically turned down all proposals from the government of Texas, and recently also those of the United States, to vacate the San Saba valley, since it serves them as winter quarters and is the only convenient route to Mexico.

The natural conditions for settlements by the whites in the basin of the Llano are more favorable since there are large arable tracts even some distance from the river.

Although the soil can be called arable, it is not by any means the heavy, black humus found in the region of New Braunfels and San Antonio de Bexar, or the fertile soil found generally in the undulating parts of Texas. It is usually light, partially sandy and will yield a good crop of corn and other grain only if fertilized and the season is not too dry. We did not visit the valley of the Concho, which next to the Llano and the San Saba, is the most important stream within the boundary of the grant. Owing to its great distance from the settlements, it would only then be of importance, if the condition of the soil warrants cultivation and the regions on the Llano and San Saba, lying between it, were first settled.

I deem it a duty toward my emigrated compatriots to make a clear statement here, according to the best of my knowledge, what chances for success the contemplated colonization of the grant, obtained by the Mainzer Verein for the protection of German immigrants, offers. After careful deliberations I must declare, although with reluctance, since so many exertions have been put forth, that the land in question on the right bank of the Colorado and the region north of the Llano is not the proper place for a settlement by the Germans, at least not at the present time. I would advance the following reasons :

1. There are no extensive areas in the whole region which are of the same fertility as in the lower accessible parts of Texas.

2. The distance from the inhabited sections of Texas is too great.

3. The Comanche Indians will become (if not dangerous) at least very annoying to any settlement north of the Llano.

With reference to the second point I would like to emphasize the fact that the price of many articles in Fredericksburg is double or even treble that of the same article when bought in Galveston or Matagorda Bay. The marketing of products will also offer great difficulties, owing to the distance. It takes loaded wagons four to five days longer to go from Fredericksburg to the nearest southern boundary of the Verein grant and this in part over rocky heights which are difficult to traverse.

The settlers will only have sufficient protection from the Indians if the government of the United States will carry out the promised military regulations for the protection of the whole boundary in Texas along the Indian territory. So far it seems no attempts have been made to carry out this plan.

The reasons advanced appear more weighty when one considers that in the most beautiful and fertile spots of western Texas, where none of the disadvantages mentioned are present, great areas of untilled soil are still to be had for a small sum. Among them I should like to mention particularly the land between New Braunfels and San Antonio; the land along the Medina and that farther west as well as east of the Colorado along the little streams such as Little River, Brushy Creek, the San Gabriel, etc.

There are also areas in the low country between the Brazos and the Colorado which, while the land is higher in price and lacks the crystal clear rivers and brooks as well as the excellent mesquite prairies of western parts of Texas, have the advantage of being closer to the coast and have natural lines of communication.

Although rather late, since many great sacrifices have already been brought, it would be to the interest of the Mainzer Verein or the society which has taken over the rights and responsibilities of the latter and in the interest of their German compatriots, to abandon the plans for colonization of the grant, and to buy suitable lands in the regions just mentioned. Then, after having first satisfied the well-founded claims of the immigrants who had priority rights, sell suitable parcels to the new arrivals at a reasonable figure.

The fact that since my departure from Texas, isolated settlements under the name of Castell and Leiningen have been founded on the Llano near the place where we crossed it, does not contradict what has been said about colonizing this region. Only if they can maintain themselves several years independently will it be established that their founding was expedient.

Chapter XXIV

Growth of New Braunfels—Washington on the Brazos
Return to Houston—Departure from Galveston—
Farewell to Texas

My traveling companions left for New Braunfels a few days after our return to Fredericksburg, while I remained in order to acquaint myself a little more with the geognostic conditions of my surroundings and to collect the numerous fossils found in the cretaceous formation.

The weather now was very disagreeable, and that to a degree hitherto not experienced by me in Texas. It was cold several days in succession, then rainy and dreary. At one time I thought I noticed snowflakes coming down with the rain. Northers had also blown during half the days of our journey, with an occasional frost at night, but the days were usually warm and pleasant. With the exception of a light shower, no rain fell during the duration of our six-week's trip. Not until March 25, when I rode out one morning to the hills which form a semicircle around Fredericksburg at a distance of two to three miles did I see the first signs of awakening spring on the slopes of the hills after passing the dormant oak forest with its blackened soil and burnt grass.

A little shrub (*Cercis reniformis Engelmann*) four to five feet in height, still leafless, was covered with pink blossoms which from a distance could be mistaken for our laurel (*Daphne Mezereum L.*) As is the case with our shrub at home, so is also this cercis in the hilly country of western Texas the harbinger of spring. Several prunus shrubs had also opened their white blossoms and on the ground a few flowers of the multicolored *Anemone Caroliniana L.* had made their appearance.

It was high time for the settlers at Fredericksburg that spring was returning and that the prairies would again be covered with their luxuriant green grass.

The greater number of the cattle and several horses had died due to insufficient food and the remainder dragged themselves around in an emaciated condition which made them unfit for work. Even my good mule, dubbed by my friends the "scientific mule," with which I had made many a strenuous journey during the past year, refused for the time to perform its duties Some imprudent individuals had lighted the grass near the hamlet in the fall of the year, and it had burned within a radius of several miles. Had the long, dry grass remained standing, the tender shoots at the base of the stalks would have been protected from the north winds and continued their growth even in winter.

It was also thought that a fresh vegetable diet would have a beneficial effect upon the health of the colonists, of whom a great number were still suffering from a form of scurvy or stomacace which spread to the lungs and usually proved fatal.

On April 13, I started on my return trip to New Braunfels in a wagon which had brought merchandise and victuals to Fredericksburg and was now returning empty. As I was passing through the long, principal street of Fredericksburg, no doubt many a one who had hoped to find in Texas the gratification of his many desires, but who instead had found in this outpost of civilization only a place offering many privations and hardships, and who probably had lost his robust health in this warm climate, looked after me with longing, since all knew that I was hurrying home to the German fatherland. I myself felt sad at heart when I saw the many neat houses, enclosed carefully with a fence, which German industry had called forth in the wilderness, far removed from the habitations of civilized human beings, and when I realized the disadvantages of the location which had been selected solely out of consideration for the future colonization of the grant lying farther north. It is my earnest desire, but I scarcely dare hope it, that the prospects for this settlement will prove less unfavorable than I view them, or may German industry and German frugality conquer these obstacles.

The farther we advanced on the romantic road from Fredericksburg to New Braunfels, the more advanced was the vegetation. The tender grass was already high enough in the valley of the Salado to furnish an abundance of food for herds of deer. On the Cibolo, the tall yucca had unfurled its white, globular blossoms.

We found several well established farms at the crossing of the Cibolo, whereas a year prior to this not a house was to be seen on the entire road from New Braunfels to San Antonio. German settlements also extended in the same manner along the Comal almost uninterruptedly for a stretch of eight miles.

After a rapid journey of three days, we arrived in New Braunfels in the evening. Much had also been accomplished here during my absence. Numerous new houses had been built and the whole place looked much more like a city. The straight streets which formerly were only indicated on the ground plan of the city, now began to appear in reality. But what was more important, in the vicinity of the city were numerous fenced-in fields which were being made ready for the planting of corn and other grain. A speculative American had laid out a new city in the bottom land between the fork of the Comal Spring and the Guadalupe within view of the city of New Braunfels, called Comaltown. He had sold lots to the immigrants, but most of them had been abandoned, due to the unhealthful location. Another city called Hortontown, had been laid out on the other bank of the Guadalupe, and although the soil was not so fertile here, several families had settled there, due to its healthful location. Several miles below New Braunfels in a pleasant and fruitful bend of the river is another small settlement, called "Frenchman's Corner." Only under the circumstance that many people, engaged in agriculture will settle in the vicinity will New Braunfels be able to maintain itself, since it is not situated on a navigable river, nor has it other commercial advantages, except that it is a central point, where the agricultural population can exchange their products for merchandise and the artisans have their permanent place of residence.

The day of my departure was near at hand and I tried to speed up the packing of my collection in order to be in New Orleans before the advent of hot weather. My preparations were all completed on April 23.

I bid a hearty farewell to my friends who had supported me so materially in the pursuit of my scientific undertaking, and who had helped to make the past year in Texas one of the most pleasurable ones of my life through their charming and instructive companionship. I boarded the stage, bound for Houston, which was waiting at the American hotel of the city. This stage had just recently been put into operation by an enterprising American. A wagon drawn by four horses goes twice a week from Houston to San Antonio and equally as often in the opposite direction. Through a branch line to Lavaca Bay, other places in the lower Guadalupe valley make connection with the main stage. The entire stretch from San Antonio to Houston is traversed in four days with regular stops for the night. The trip from New Braunfels to Houston requires three and one-half days and the fare, including free transportation for the trunk, is twenty dollars. I contemplated for the last time with keen interest the houses of the city, the majority of which I had seen under construction. May they become the habitations of a contented and happy people, since their builders had to endure so many difficulties and hardships, as well as witness so much human misery.

The last houses on the long, principal street lay behind us and we passed a large log house, belonging to an acquaintance (Mr. W. from Hanover) where I had visited often and where the road led to the ford over the Guadalupe and to the incomparable, beautiful confluence of the Comal and Guadalupe, with their wonderful clear water, and overhanging, dainty foliated cypress trees. A few moments later we had passed through the ford over the Guadalupe.

For the last time I contemplated the white limestone layers forming its bed, where I had spent many an evening before sundown, enjoying the cool breezes emanating from

the water and gathering fossils. Now we drove into the
beautiful stream itself, whose clear water flows tempestuously over the rocky bed. I had never seen its water so low.
At this same time in the previous year it was a raging torrent,
almost overflowing its banks, carrying with it tree trunks in
great numbers. During the past winter, however, the ferry
was hardly used at all, and mounted men and vehicles had
used the ford almost exclusively. The low stand of the water
continued even now, since it had rained very little for the
past six months. Upon arriving on the other side, I looked
involuntarily back upon the city. Only when the Verein
building, lying on a hill, with its projecting shingled roof
had disappeared from view, did I turn my attention to my
immediate surroundings. I first inspected the vehicle a little
closer, to which I had entrusted my person and my trunk.
The result of my inspection was not very gratifying. It was
the framework of a carriage on which a square box without
springs was fastened. At first I sat in a seat suspended on
straps, but when they broke, after we had scarcely left New
Braunfels, I was obliged to sit on my trunk. The coachman,
a tall, lean person with a serious, firm expression of countenance and ruddy face, was also conductor, as is the custom
in America. He presented an unmistakable specimen of his
nation as well as his trade, in his neat suit and white shirt
sleeves and also in the safety with which he handled his four
black mules, distinguishing himself just as materially from
our German postilion as the American farmer does from a
peasant from Mecklenburg.

This coachman was, moreover, a well-traveled man, as I
found out during the course of our conversation. He had
driven a diligence (large stagecoach) a number of years
between the cities of Vera Cruz and the capital of Mexico,
where all coachmen are Americans, since the Mexicans all
know how to drive mules and to ride, but not to drive vehicles.

We came to Seguin toward sundown where I felt myself
transported suddenly to an American manner of living. A

number of tall, haggard figures sat in complete silence on the veranda of the frame tavern, rocking to and fro on their chairs. Their feet were propped against the pillars on the veranda, on the same level with their body, and they were expectorating to the right and to the left, eyeing us apparently with an air of indifference, but at the same time scrutinizing the stranger very minutely. To one side was the inevitable apparatus for washing, as well as for drinking water, which every stranger uses upon his arrival. It was composed of a wooden pail, with a gourd serving as a dipper, and a tin washbasin.

Soon a gong sounded which called us to supper. This was served in a room in the rear of the building and consisted of the well-known victuals of a Texas meal, i.e., fried bacon, cornbread and coffee. In addition to this, little hot wheat biscuits and fresh butter were passed around. These two articles are more or less a luxury and are not found in every farmhouse. Barefooted negro girls waited on us and fanned the flies away with long, green branches. If they were inattentive in their serving, a matron, sitting at the end of the table, serving the coffee with great dignity, reminded them of their duty with words of flattery, or by pinching their ears rather roughly or by slapping them on the cheeks. Shortly after supper I withdrew to the sleeping quarters assigned to the male guests, where four large beds were standing, but where it was self-evident that each bed would be occupied by two persons.

We got an early start on the following morning and arrived at Gonzales in the afternoon, where we took our second night's lodging. This place, established under Mexican rule, did not show any signs of growth. In fact everything looked desolate and dilapidated. Gonzales is a forlorn, unhealthy place, as can be expected, owing to its location near the extensive bottom of the Guadalupe. Also here many persons had died during the previous summer and the remaining ones looked as though they had just arisen from a sickbed, and had been fed on calomel. The only persons who seemed to prosper were the physician who was at the same time druggist and postmaster and the proprietors of the various barrooms or "groceries," i.e., merchandise stores, as they were euphoniously called.

Although I found the lodging in the home of the physician clean and tidy, nevertheless I was glad when I left this dismal place in the morning and entered the forested bottom of Peach Creek, where everything was decorated in verdant green. After crossing this creek, which was entirely dry, but is often a raging torrent during certain times of the year which detains traffic for several days, we entered the monotonous oak forest, which extends without interruption to the Colorado. Toward noon we came to La Grange, the beautiful little city lying in the bend of the Colorado at the foot of a hill with precipitous slopes which presents a most pleasant view with its white frame buildings.

The number of our passengers was augmented here. A wagon which had left Austin at the same time we left New Braunfels met us here. The combined passengers and baggage were transferred to this wagon, thus creating a very crowded condition; nevertheless everyone was glad to get the opportunity to leave, since extra coaches, which are also unknown throughout America, were not to be found here. The passengers now were composed in addition to myself, of the following persons: An old, wealthy merchant of Irish descent from San Antonio who was taking his seven-year-old son to a Catholic educational institution in New Orleans; a well-educated young lawyer, who had attended the sessions of the Supreme Court in Austin and who was now returning to his home in Galveston; a talkative old Catholic priest of the Jesuit College in St. Louis, dressed in his black, clerical garb, who was returning to St. Louis by way of New Orleans. He had just made his annual missionary journey through Texas, as he told me. In the course of our conversation he often took occasion to praise the excellency of the free American institutions. However, I was inclined to doubt the sincerity of his admiration for them, for in the course of

our conversation, I found him unfavorably inclined toward
freedom in the intellectual domain of scientific investigations; although it was known to me that this Jesuit college
in St. Louis, as well as others in other parts of the Union vie
with other corporations in celebrating with speeches and
public entertainments, the great national festivals, such as
the Fourth of July and Washington's birthday, in order to
show their nationalism and respect for the free institutions
of the land.

There was furthermore a young German who was a druggist by profession, but who had operated a saloon in New
Braunfels during the past year. Since he had very much
competition and due to the fact that several of his best cus-
tomers had left suddenly to fight in the defense of the new
fatherland against Mexico, without first bidding him farewell, his business had not been very lucrative; therefore he
had abandoned this thankless occupation and now had the
intention of becoming a physician among the Americans.
In his company was a young Prussian nobleman who, although scarcely six months in Texas, had already farmed
on the Colorado near Columbus, and later had formed a
partnership with two other men in Fredericksburg. After
retailing a few bottles of whiskey, he appeared to have concluded that Texas was not the land to gratify his desires, and
he was therefore hurrying back to Europe in order to relate
his experiences to his relatives, who, as he stated, would be
delighted at his return.

Then there sat near me a German woman of the middle
class who had just recently obtained a divorce from her husband, living in New Braunfels. She was returning to Europe.
A merchant from Austin, traveling to Houston to buy merchandise, completed the number of passengers in addition
to the coachman. Evidently this woman had been entrusted
to his care and with true American gallantry he was very attentive to her, and she in turn was Americanized enough to
accept these attentions with a most condescending air, as if
he were merely doing his duty.

The shortest route from La Grange to Houston goes over San Felipe de Austin through a very fertile region, where near the hamlet Industry on Mill Creek many Germans had settled ten or twelve years prior to this time. However, we did not take this route but made a detour to Washington, due to the mail, which according to a contract with the Postmaster General at Washington, was carried by stagecoach.

We arrived at the latter place in the afternoon. It was the most wretched and forlorn place of all the so-called cities of Texas. Washington had been for a long time the seat of government for the Republic of Texas and as a consequence of this a number of large houses had been built. Several industries had also been established, so that the population had increased to one thousand. Subsequently the seat of government had been transferred to Austin and the doom of Washington was sealed. It does not require a long time in Texas to make the decline of a place very apparent, owing to the light construction of the buildings. And I believe if the inhabitants of Austin would abandon that city, it would be difficult to determine after a lapse of ten years where the city had stood.

This Washington in Texas had not quite reached such a point, but it was rapidly nearing that stage. We saw several large, vacant houses with broken window panes, missing shingles and loose boards. Those occupied also seemed to lack the care necessary for their preservation. The former saloon-keeper from New Braunfels, who had selected this place to make his debut as physician, remained here, and judging from the experience of former years, he had not made a bad choice, for there was a current saying, that a representative to Congress, who had braved the fever of Washington one year, showed just as much courage for the Republic as another who had participated in a campaign against the Mexicans.

We crossed the Brazos over a ferry not far from Washington and came to its broad forested bottom. The road was bottomless and we made very slow progress. At the same time a long-threatening, real Texas thunderstorm, burst over us. Since the wagon was uncovered, I sought to protect myself as well as possible with my cloak and a buffalo robe, but to no avail, for the attentive merchant from Austin held a large umbrella over his "lady" in such a manner, that the water fell like a cascade upon unfortunate me. All remonstrances, even calling attention to my fever-ridden conditions, would have been to no avail when it involved "protecting a lady" and I resigned myself to the inevitable with Christian fortitude.

The demand of the driver that all male passengers alight and walk, since the mud was too deep to permit a loaded coach to pass, therefore had no terrors for me. We waded in this black mud, a foot deep, for about a half hour, when the driver declared that the road was becoming firm again, and ordered us to climb into the coach. In the meantime it had become dark, but despite this, our coachman drove as fast as his horses could go in order to make up for lost time. The result was that while crossing a little boggy creek, the wheels on the one side of the wagon slipped off the saplings which had been placed there in the absence of a bridge, and now our wagon came to a standstill. The entire stage had to be unloaded, and since the exhausted horses were not able to pull it out of the bog, we had to get help from a plantation several miles distant. After a long wait, the owner of the farm came with a half dozen slaves, who dislodged the wagon easily. We arrived late at the station where we were to stop for the night. It was a large, stately manor, in which the promoter of the whole stagecoach system of Texas lived. He tried to make us forget the difficulties we had encountered through his obligingness and friendliness, and assured us that good covered coaches for the entire system had been ordered from the northern states. This was good news for me—at least in the interest of those who would follow after me. Unfortunately I did not find a good night's rest here, although I longed for it. A number of beds had been prepared on the floor of the attic of the house for this large company and here everyone was expected to find a place to sleep. I had just captured a small mattress for my own use and had spread a thin cotton blanket over my chilly frame, when my host assigned another individual, utterly unknown to me to share my bed and thin blanket with me. The results were that one side of my body was uncovered during the entire night and I was compelled to tug on the blanket in order to prevent the tendency of my snoring friend, egotistical even in his sleep, to expose also the other side of my body.

However, the difficulties of the journey were over, for on the next morning after a good breakfast, we climbed into an excellent red-painted "Troy Coach," such as are manufactured in Troy on the Hudson and used throughout the United States. In it we traveled without difficulty the remaining fifty miles to Houston over the monotonous, treeless Houston Prairie, whose level surface was unbroken even by the slightest elevation. We arrived there late in the evening. After having gone here and there on the extreme outposts of civilization for over a year, the city with its spacious hotel, the Houston House, its brightly illumined decorated barrooms, and various billiard halls, appeared very grand and magnificent to me.

Daylight, however, broke this magic spell, for I soon convinced myself that the city was essentially the same as it had been the previous year, although new and pretty homes and several stores filled with merchandise, had been added to it. I had to wait three days here for a steamboat to Galveston. Because of the war with Mexico, several steamboats sailing regularly between Houston and Galveston, had been requisitioned by the government of the United States to be used as transports for troops and provisions to the Rio Grande, and they are now being used on the Rio Grande itself.

On this trip I saw the stretch on the narrow Buffalo Bayou which on my first journey I had passed during the night. It is a peculiar sight to sail on this narrow river, whose width in some places is taken up almost entirely by a comparatively large boat, so that the trees and shrubs on the twenty to thirty feet high banks can almost be touched from its deck.

Turning a boat around or passing another is all out of the question. The magnolia trees which occur frequently on the banks of Buffalo Bayou, had partially unfolded their white blossoms as large as plates and everywhere the luxuriant vegetation of a Texas spring was visible.

We landed in Galveston after a journey lasting twelve hours. On this occasion I took lodging in the German hotel of Mr. Beissner, where I spent eight comfortable, pleasant days in the company of other cultured Germans. The city had been enlarged and beautified considerably during my absence. Among the newly erected public buildings I noticed a new marketplace without which of course no American city can do. However, the activity and trade of the previous year was not in evidence. The streets seemed deserted; many stores were closed; and in the remaining ones the clerks stood around idly in the doorway. In the harbor, which in the previous year was filled with sail and steamboats, only an American brig and a few coasting vessels were anchored. The large American hotel, the "Tremont House," where at the time of my arrival eighty to one hundred persons gathered daily for their meals, was on the verge of closing for lack of patronage.

Several unfavorable conditions had combined to bring about this unfavorable change. On one hand, the cotton crop of the previous year had been destroyed almost entirely by insects in a great portion of Texas. With the falling off of the export of this excellent staple article of the country, which was regularly and almost exclusively sent by way of Galveston, the merchants were deprived of a great part of their income. On the other hand, this condition also reacted unfavorably upon the commercial relations, since the planters, who had also experienced a considerable falling off in their income, restricted their purchase to the most necessary articles.

The fact that regular communication with New Orleans was interrupted, also acted unfavorably upon commerce. Formerly two steamships, built especially for this purpose, sailed regularly between these two places, arriving weekly. Shortly after the outbreak of the war with Mexico, both ships had been leased by the government of the United States for military purposes. Since then communication was maintained only through a small steamboat and a schooner, which curtailed passenger trade considerably, and caused such irregularity, that the mail from New Orleans was often delayed three weeks.

Commerce and activity in the city had finally also received a setback in consequence of the sad reports which the German immigrants (who had arrived here in the fall and winter and had been retained at the coast until the following summer) had sent home. In this and the previous year, hardly any immigrants had arrived.

The city with its cheerful white houses and pretty gardens, in which roses and oleanders were in full bloom, pleased me still more than on my first visit. Despite the fact that it lies upon a narrow, treeless, sandy island, scarcely above the level of the sea, where nearly all the necessities of life, such as grain, wood and even water must be brought over from the mainland or New Orleans, one can live comfortably and cozily here. On the same day of my arrival I visited the beautiful beach on the seaward side of the island. It had become a more universal custom to bathe here and no more suitable place could have been found anywhere.

The long-expected steamboat *Yacht* finally arrived May 7, 1847, and on the following day I boarded it, bound for New Orleans. When the stronger rocking of the boat indicated that we had passed the harbor bar, and when soon thereafter the land of the narrow island appeared only as a low streak, I felt that it was time to say farewell to Texas.

During my stay of more than a year, I had developed interest and love for the beautiful land of meadows which faces a bright future; and it filled my heart with sadness to be compelled to bid it farewell forever. However there remain with me agreeable and rich memories and I will always follow from the distance the further development of this country with keen interest. May its wide, green prairies become the home of a large and happy population.

INDEX OF PLACE NAMES & WATERCOURSES

INDEX OF SCIENTIFIC NAMES